T0290040

Actuarial Loss Models

Actuarial loss models are statistical models used by insurance companies to estimate the frequency and severity of future losses, set premiums, and reserve funds to cover potential claims. Actuarial loss models are a subject in actuarial mathematics that focus on the pricing and reserving for short-term coverages.

This is a concise textbook written for undergraduate students majoring in actuarial science who wish to learn the basics of actuarial loss models. This book can be used as a textbook for a one-semester course on actuarial loss models. The prerequisite for this book is a first course on calculus. The reader is supposed to be familiar with differentiation and integration.

This book covers part of the learning outcomes of the Fundamentals of Actuarial Mathematics (FAM) exam and the Advanced Short-Term Actuarial Mathematics (ASTAM) exam administered by the Society of Actuaries. It can be used by actuarial students and practitioners who prepare for the aforementioned actuarial exams.

Key Features:

- Review core concepts in probability theory.
- Cover important topics in actuarial loss models.
- Include worked examples.
- Provide both theoretical and numerical exercises.
- Include solutions of selected exercises.

Guojun Gan is an Associate Professor in the Department of Mathematics at the University of Connecticut, Storrs, Connecticut, USA. He received a BS degree from Jilin University, Changchun, China, in 2001 and MS and PhD degrees from York University, Toronto, Canada, in 2003 and 2007, respectively. His research interests are in the interdisciplinary areas of actuarial science and data science.

CHAPMAN & HALL/CRC Series in Actuarial Science

Series Editors
Arthur Charpentier, Edward W. Frees, Montserrat Guillen, Steven Haberman and Greg Taylor

Recently Published Titles

Actuarial Loss Models
A Concise Introduction
Guojun Gan

For more information about this series, please visit:
https://www.routledge.com/Chapman--HallCRC-Series-in-Actuarial-Science/book-series/CHCRCACTSCI

Actuarial Loss Models
A Concise Introduction

Guojun Gan

CRC Press is an imprint of the
Taylor & Francis Group, an **informa** business
A CHAPMAN & HALL BOOK

Designed cover image: © 2025 Guojun Gan

First edition published 2025
by CRC Press
2385 NW Executive Center Drive, Suite 320, Boca Raton FL 33431

and by CRC Press
4 Park Square, Milton Park, Abingdon, Oxon, OX14 4RN

CRC Press is an imprint of Taylor & Francis Group, LLC

Library of Congress Cataloging-in-Publication Data

Names: Gan, Guojun, 1979- author.
Title: Actuarial loss models : a concise introduction / Guojun Gan.
Description: First edition. | Boca Raton, FL : CRC Press, 2025. | Series: Chapman & Hall/CRC series in actuarial science | Includes bibliographical references and index.
Identifiers: LCCN 2024016620 (print) | LCCN 2024016621 (ebook) | ISBN 9781032777658 (hardback) | ISBN 9781032778082 (paperback) | ISBN 9781003484899 (ebook)
Subjects: LCSH: Insurance--Statistical methods. | Loss ratios (Insurance)
Classification: LCC HG8781 .G36 2025 (print) | LCC HG8781 (ebook) | DDC 368/.01--dc23/eng/20240419
LC record available at https://lccn.loc.gov/2024016620
LC ebook record available at https://lccn.loc.gov/2024016621

ISBN: 978-1-032-77765-8 (hbk)
ISBN: 978-1-032-77808-2 (pbk)
ISBN: 978-1-003-48489-9 (ebk)

DOI: 10.1201/9781003484899

Typeset in CMR10
by KnowledgeWorks Global Ltd.

Publisher's note: This book has been prepared from camera-ready copy provided by the authors.

To my students

Contents

Preface

Actuarial loss models are statistical models used by insurance companies to estimate and predict insurance losses in the future. Unlike non-insurance products whose costs are usually known in advance; insurance is a promise to pay something in the future if certain events happen during a specified time period. As a result, actuarial loss models are essential for insurance companies in pricing, reserving, risk management, and financial planning.

This is a textbook written for undergraduate students majoring in actuarial science who wish to learn the basics of actuarial loss models. This book can be used as a textbook for a one-semester course on actuarial loss models. The prerequisite for this book is a first course in calculus. The reader is supposed to be familiar with differentiation and integration.

This book covers part of the learning outcomes of the Fundamentals of Actuarial Mathematics (FAM) exam and the Advanced Short-Term Actuarial Mathematics (ASTAM) exam administered by the Society of Actuaries. It can be used by actuarial students and practitioners who prepare for the aforementioned actuarial exams.

This book contains exercises at the end of each section to help students learn the concepts and facts introduced in the section. Some exercises are theoretical and involve proving some statements. Other exercises are numerical and involve computing some quantities. Many of the numerical exercises are adopted from the sample questions of the actuarial exams administered by the Society of Actuaries. For example, many exercises in the first chapter are adopted from sample questions of the Probability (P) exam. Many exercises in other chapters are adopted from the FAM and the ASTAM exams.

A key feature of this book is that it provides detailed solutions for selected exercises. For theoretical exercises, the solutions include detailed proofs. For numerical exercises, the solutions include detailed calculations. Unlike the sample solutions provided by the Society of Actuaries, the solutions given in this book connect the underlying problems to the theory introduced in the book. The detailed solutions will help students to learn the materials.

This book also includes some commonly used results from calculus and special functions in the appendices. For example, properties of derivatives and integration by parts are included. Students who do not use these results may forget the details. Definitions and properties of special functions such as the gamma function and the beta function are also provided.

This book grew out of my lecture notes for the course "Math 3639: Actuarial Loss Models," which I taught at the University of Connecticut. In writing a book on actuarial loss models, one is bound to be influenced by the classics *Loss Models: From Data to Decisions* by Klugman, Panjer, and Willmot and *Nonlife Actuarial Models: Theory, Methods, and Evaluation* by Tse. I thank the authors for writing such comprehensive textbooks on actuarial loss models. I would also like to take this opportunity to express my thanks to my students, friends, and colleagues from the University of Connecticut who have read and provided valuable feedback on the draft of this book.

Storrs, CT
Guojun Gan
March 2024

1

Probability Theory

Probability theory plays a fundamental role in modeling actuarial losses. In this chapter, we present various concepts and facts of probability theory that are useful in creating, estimating, and communicating actuarial loss models.

1.1 Probability Spaces

A probability space consists of three components: a sample space, an event space, and a probability measure (see Definition 1.1). Rigorous treatment of probability theory requires measure theory (see, for example, [20]).

Definition 1.1 (Probability space)**.** A probability space is a mathematical construct that models a random experiment. A probability space is denoted by a triple $(\Omega, \mathscr{F}, P)$, where Ω is the set of all possible outcomes of the random experiment, $\mathscr{F}$ is a set whose elements are subsets of Ω, and $P : \mathscr{F} \to [0, 1]$ is a probability measure.

In Definition 1.1, the event space $\mathscr{F}$ is a set of events, which are subsets of the sample space Ω. The event space $\mathscr{F}$ satisfies the following properties:

(a) $\Omega \in \mathscr{F}$;

(b) $\mathscr{F}$ is closed under complementation. That is, if $A \in \mathscr{F}$, then $A^c \in \mathscr{F}$, where $A^c = \Omega / A$;

(c) $\mathscr{F}$ is closed under countable unions. That is, if $A_1, A_2, \ldots \in \mathscr{F}$, then

$$\bigcup_{i=1}^{\infty} A_i \in \mathscr{F}.$$

A set that satisfies the above three properties is called a σ-algebra. It is a concept from measure theory. It is clear that the event space always contains $\emptyset$ and Ω. Here $\emptyset$ denotes the empty set. Here we used the concept of countable. A set is said to be countable if it has a finite number of elements or its elements

DOI: 10.1201/9781003484899-1

can be numbered by natural numbers. Similarly, countable unions means a finite number of union operations or the union operations can be numbered by natural numbers.

If Ω is countable, the event space can be made to be the set that contains all subsets of Ω. However, when Ω is uncountable (e.g., the interval $[0,1]$), the event space cannot be the set that contains all subsets of Ω.

Probability measures are functions mapping from the event space to the real numbers. Definition 1.2 lists the three probability axioms, which are conditions of probability measures.

Definition 1.2 (Probability measure). A probability measure P is a function from an event space $\mathscr{F}$ to the real numbers $\mathbb{R}$ that satisfies the following properties:

(a) $P(A) \geq 0$ for all $A \in \mathscr{F}$;

(b) $P(\Omega) = 1$;

(c) If $\{A_1, A_2, \ldots\}$ is a countable collection of disjoint events, then

$$P\left(\bigcup_{i=1}^{\infty} A_i\right) = \sum_{i=1}^{\infty} P(A_i).$$

The above three properties are called the axioms of probability.

Examples 1.1 and 1.2 give two examples of probability spaces. In Example 1.1, the sample space is finite. In Example 1.2, the sample space is countable If a sample space is countable, the event space can be the power set of the sample space.

Example 1.1. Consider a random experiment of rolling a fair dice, whose faces are labeled by six digits 1, 2, …, 6. Determine the corresponding probability space $(\Omega, \mathscr{F}, P)$ and calculate the probability of the event $\{1,2\}$.

Solution. For this random experiment, the sample space is

$$\Omega = \{1,2,3,4,5,6\}.$$

The event space will consist of all subsets of Ω, i.e.,

$$\mathscr{F} = 2^{\Omega} = \{\text{all subsets of } \Omega\}.$$

Here 2^{Ω} denote the power set of Ω that contains all subsets of Ω. Since the coin is a fair coin, the probability of getting any face is the same, i.e.,

$$P(\{i\}) = \frac{1}{6}, \quad i = 1,2,\ldots,6.$$

From the third axiom, we can calculate the probability of the event $\{1, 2\}$ as follows:

$$P(\{1,2\}) = P(\{1\}) + P(\{2\}) = \frac{1}{3}.$$

□

Example 1.2. Three persons A, B, and C take turns to flip a fair coin. Person A flips the coin first, then B, then C, then A, and so on and so forth. The first person to get a head wins. Determine the corresponding probability space and calculate the probability of the event $\{A \text{ wins}\}$.

Solution. For this random experiment, the sample space contains an infinity number of outcomes:

$$\Omega = \{H, TH, TTH, TTTH, \ldots, TTT\cdots\},$$

where H denotes a head outcome and T denotes a tail outcome. The event space $\mathscr{F}$ is the power set of the sample space.

The event $\{A \text{ wins}\}$ contains the following outcomes:

$$\{A \text{ wins}\} = \{H, TTTH, TTTTTTH, \ldots\}.$$

The probability of an individual outcome can be calculated by the product of the probabilities of heads and tails:

$$P\left(\underbrace{T\cdots T}_{m}H\right) = \frac{1}{2^{m+1}}, \quad m = 0, 1, \ldots.$$

By the third axiom, the probability of this event can be calculated as follows:

$$\begin{aligned} P(\{A \text{ wins}\}) &= \sum_{k=0}^{\infty} P\left(\underbrace{T\cdots T}_{3k}H\right) = \sum_{k=0}^{\infty} \frac{1}{2^{3k+1}} \\ &= \frac{1}{2}\sum_{k=0}^{\infty} \frac{1}{8^k} = \frac{1}{2}\cdot\frac{1}{1-\frac{1}{8}} = \frac{4}{7}. \end{aligned}$$

□

***

Exercise 1.1. Let $(\Omega, \mathscr{F}, P)$ be a probability space. Let $E, F \in \mathscr{F}$. Show that

$$P(E \cup F) = P(E) + P(F) - P(E \cap F).$$

Exercise 1.2. Let A, B, and C be three events. Show that

$$\begin{aligned} P(A \cup B \cup C) =& P(A) + P(B) + P(C) - P(A \cap B) - P(B \cap C) - \\ & P(C \cap A) + P(A \cap B \cap C). \end{aligned}$$

Exercise 1.3. Let E and F be two events such that $E \subset F$. Show that

$$P(F) \geq P(E).$$

Exercise 1.4 (Bonferroni's inequality)**.** Let E and F be two events. Show that

$$P(E \cap F) \geq P(E) + P(F) - 1.$$

Exercise 1.5. Three persons A, B, and C take turns to flip a fair coin. Person A flips the coin first, then B, then C, then A, and so on and so forth. The first person to get a head wins. Calculate the probabilities of the events $\{B \text{ wins}\}$ and $\{C \text{ wins}\}$.

Exercise 1.6. Let A and B be two events. Suppose that $P(A \cup B) = 0.7$ and $P(A \cup B^c) = 0.9$. Calculate $P(A)$.

Exercise 1.7. After surveying a large group of patients recovering from shoulder injuries, an actuary found that:

(a) 22% visit both a physical therapist and a chiropractor.

(b) 12% visit neither a physical therapist nor a chiropractor.

(c) The probability that a patient visits a chiropractor exceeds by 0.14 the probability that a patient visits a physical therapist.

Calculate the probability that a randomly chosen patient in this group visits a physical therapist.

Exercise 1.8. The health plan of a large company has three supplementary coverage options A, B, and C. The plan includes a provision that an individual employee may choose exactly two supplementary coverage or none. The proportions of the company's employees that choose A, B, and C are 1/4, 1/3, and 5/12, respectively. Calculate the probability that a randomly selected employee will choose no supplementary coverage.

---***

1.2 Conditional Probability and Independence

Conditional probability is an important concept in probability theory. Conditional probability, as its name indicates, refers to probability given some partial information. Definition 1.3 gives the definition of conditional probabilities.

Definition 1.3 (Conditional probability)**.** Let E and F be two events. Suppose that $P(F) > 0$. Then, the conditional probability that E occurs given that F has occurred is defined as:

$$P(E|F) = \frac{P(E \cap F)}{P(F)}.$$

From the definition of the conditional probability, we see that

$$P(E|F) \geq \frac{P(E \cap F)}{1} = P(E \cap F).$$

If $P(F) < 1$, the above inequality is strict, i.e., $P(E|F) > P(E \cap F)$. Example 1.3 illustrates this.

Example 1.3. Consider a random experiment of rolling a fair dice, whose faces are labeled by numbers from 1 to 6. Calculate the probability that the number 1 or 3 is rolled given that the number rolled is odd.

Solution. This is an example of calculating conditional probabilities. Let F be the event that an odd number is rolled. Let E be the event that the number 1 or 3 is rolled. Then $F = \{1, 3, 5\}$ and $E = \{1, 3\}$. Hence

$$P(E|F) = \frac{P(E \cap F)}{P(F)} = \frac{P(\{1,3\} \cap \{1,3,5\})}{P(\{1,3,5\})} = \frac{P(\{1,3\})}{P(\{1,3,5\})} = \frac{\frac{2}{6}}{\frac{3}{6}} = \frac{2}{3}.$$

Without the information that an odd number is rolled, the probability of the number 1 or 3 is rolled is

$$P(\{1,3\}) = \frac{2}{6} = \frac{1}{3},$$

which is lower than the conditional probability we just calculated. □

Independence is another important concept in probability theory. Definition 1.4 gives the definition of independence of events. It is worth noting that the independence of more than two events requires events in every subset to be independent.

Definition 1.4 (Independent events)**.** Two events E and F are said to be independent if

$$P(E \cap F) = P(E)P(F).$$

If two events are not independent, they are said to be dependent.

A finite number of events E_1, E_2, …, E_n are said to be independent if for any $r \leq n$ and any subset of indices $1 \leq i_1 < i_2 < \cdots < i_r \leq n$,

$$P(E_{i_1} \cap E_{i_2} \cap \cdots \cap E_{i_r}) = P(E_{i_1})P(E_{i_2}) \cdots P(E_{i_r}).$$

An infinite number of events are said to be independent if events in every finite subset are independent.

From Definitions 1.3 and 1.4, we see that if two events E and F are independent, then

$$P(E|F) = \frac{P(E \cap F)}{P(F)} = \frac{P(E)P(F)}{P(F)} = P(E)$$

and

$$P(F|E) = \frac{P(F \cap E)}{P(E)} = \frac{P(F)P(E)}{P(E)} = P(F).$$

In other words, if E and F are independent, then the conditional probability of one given another is the same as the unconditional probability.

***

Exercise 1.9 (Bayes' theorem). Let E and F be two events such that $P(E) > 0$ and $P(F) > 0$. Show that

$$P(E|F) = \frac{P(F|E)P(E)}{P(F)}.$$

Exercise 1.10. Let E and F be two independent events. Show that E and F^c are also independent.

Exercise 1.11. Let E and F be two events. Suppose that $0 < P(F) < 1$. Show that

$$P(E) = P(E|F)P(F) + P(E|F^c)P(F^c).$$

Exercise 1.12 (Multiplication rule). Let E_1, E_2, …, E_n be events such that $P(E_1 \cap E_2 \cap \cdots \cap E_n) > 0$. Show that

$$P(E_1 \cap E_2 \cap \cdots \cap E_n) = P(E_1)P(E_2|E_1) \cdots P(E_n|E_1 \cap E_2 \cap \cdots \cap E_{n-1}).$$

Exercise 1.13. After studying the insurance preferences of a large group of automobile owners, an actuary found that:

(a) 15% purchase both collision and disability coverages.

(b) The probability that an automobile owner purchases collision coverage is twice the probability of purchasing disability coverage.

(c) Purchasing collision coverage is independent of purchasing disability coverage.

Calculate the probability that a randomly selected automobile owner from this group purchases neither collision nor disability coverage.

Exercise 1.14. After studying the prevalence of three health risk factors – high blood pressure, smoking, and high blood cholesterol – within a large group of women, an actuary found that:

(a) For each of the three factors, 10% in the group have only this risk factor and no others.

(b) For any two of the three factors, 12% in the group have exactly these two risk factors.

(c) The probability that a woman has all three risk factors, given that she has high blood pressure and smokes, is 1/3.

Calculate the probability that a woman has none of the three risk factors, given that she does not have high blood pressure.

***

1.3 Random Variables

Random variables are quantities of interest when a random experiment is performed. For example, the number of heads when a coin is flipped five times is a random variable. In fact, random variables are functions that map from a sample space into the real numbers (see Definition 1.5).

Definition 1.5 (Random variable). Let $(\Omega, \mathscr{F}, P)$ be a probability space. A random variable X on Ω is a mapping from Ω into the real numbers such that $\{X \le x\} \in \mathscr{F}$ for every real number x.

Here $\{X \le x\}$ is a subset of Ω. In particular, it means the following:

$$\{X \le x\} = \{\omega \in \Omega : X(\omega) \le x\}.$$

A random variable is a mapping from the sample space to the real numbers. However, not every mapping from the sample space to the real numbers is a random variable. Definition 1.5 imposes a condition on mappings. The condition is that the event $\{X \le x\}$ should belong to the event space $\mathscr{F}$ for every real number x. If an event $\{X \le x\}$ is not in $\mathscr{F}$ for some x, then its probability is not defined. The reason is that the probability measure P is only defined for events in $\mathscr{F}$.

Example 1.4. Consider a random experiment of tossing three fair coins. Let Y denote the number of heads that appear. Determine whether Y is a random variable.

Solution. The mapping Y is a random variable. The sample space of this random experiment is

$$\Omega = \{HHH, HHT, HTH, HTT, THH, THT, TTH, TTT\}.$$

There are eight different outcomes. The event space $\mathscr{F}$ is the power set of Ω.

The value of Y is given by

$$Y(\omega) = \begin{cases} 0, & \text{if } \omega = TTT, \\ 1, & \text{if } \omega \in \{HTT, THT, TTH\}, \\ 2, & \text{if } \omega \in \{HHT, HTH, THH\}, \\ 3, & \text{if } \omega = HHH. \end{cases}$$

It is clear that $\{Y \leq x\} \in \mathscr{F}$ for every real number x. For example, $\{Y \leq -1\} = \emptyset \in \mathscr{F}$. □

For a random variable, we are interested in its values that have positive probabilities around them. These values form a set called the support of the random variable. Definition 1.6 gives the mathematical definition.

Definition 1.6 (Support)**.** The support of a random variable X is the set of all possible values that X can take and is denoted by supp(X), i.e.,

$$\text{supp}(X) = \{x \in \mathbb{R} : P(X \in (x - r, x + r)) > 0 \text{ for all } r > 0\}.$$

Definition 1.7 defines the cumulative distribution function of a random variable. According to the definition, we see that a cdf $F(x)$ has the following properties:

(a) $F(x) \in [0, 1]$ for all $x \in (-\infty, \infty)$.

(b) $F(x)$ is nondecreasing.

(c) $F(x)$ is right-continuous, i.e., for every $x \in (-\infty, \infty)$,

$$\lim_{h \downarrow 0} F(x + h) = F(x),$$

where $h \downarrow 0$ means that h approaches 0 from the right hand side.

(d) $\lim_{x \to -\infty} F(x) = 0$ and $\lim_{x \to \infty} F(x) = 1$.

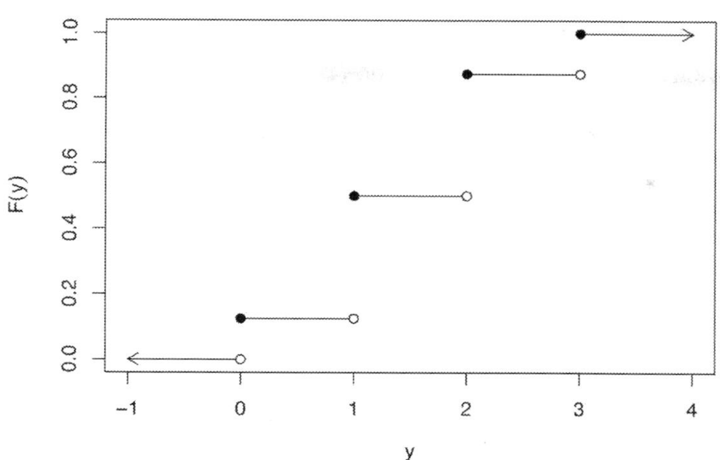

FIGURE 1.1
The cdf of the random variable Y given in Example 1.5.

Definition 1.7 (Cumulative distribution function)**.** The cumulative distribution function (cdf), also called the distribution function, of a random variable X is defined as

$$F_X(x) = P(X \leq x), \quad -\infty < x < \infty,$$

where $P(X \leq x)$ denotes the probability of the event $\{X \leq x\}$.

Example 1.5 gives the cdf of the random variable from Example 1.4. For discrete random variables, their cdfs are step functions. Figure 1.1 shows the cdf.

Example 1.5. Consider the random variable Y given in Example 1.4. Determine its cdf.

Solution. Since Y is a discrete random variable, its cdf is a step function:

$$F(y) = \begin{cases} 0, & \text{if } y < 0, \\ \dfrac{1}{8}, & \text{if } y \in [0, 1), \\ \dfrac{1}{2}, & \text{if } y \in [1, 2), \\ \dfrac{7}{8}, & \text{if } y \in [2, 3), \\ 1, & \text{if } y \geq 3. \end{cases}$$

□

A function that is related to the cdf is the survival function, which is defined in Definition 1.8.

Definition 1.8 (Survival function)**.** The survival function of a random variable X is defined as

$$S_X(x) = 1 - F_X(x),$$

where $F_X(x)$ is the cdf of X.

Definition 1.9 gives the classification of different types of random variables. For a discrete random variable, its support is just the set of values that have positive probabilities. That is, if X is discrete, then its support is

$$\text{supp}(X) = \{x \in \mathbb{R} : P(X = x) > 0\}.$$

Definition 1.9 (Types of random variables)**.** A random variable is called discrete if its support is countable. A random variable is called continuous if its cdf is continuous and differentiable everywhere except for a countable number of values. A random variable is usually called mixed if it is neither discrete nor continuous.

If the cdf of a random variable is differentiable, we can define the probability density function as given in Definition 1.10. For continuous random variables, we usually can do this. For discrete random variables, their cdfs are step functions, which are not differentiable. They do not have density functions. However, we can define the probability mass functions for discrete random variables (see Definition 1.11).

Definition 1.10 (Probability density function)**.** The probability density function (pdf), also called the density function, of a random variable X is defined as the derivative of its cdf, i.e.,

$$f(x) = F'(x) = -S'(x),$$

where $F(x)$ and $S(x)$ are the cdf and the survival function of X, respectively. Hence

$$F(x) = \int_{-\infty}^{x} f(s)\,\mathrm{d}\,s, \quad S(s) = 1 - \int_{-\infty}^{x} f(s)\,\mathrm{d}\,s.$$

Definition 1.11 (Probability function). The probability function (pf), also called the probability mass function, of a random variable X is defined as

$$p_x = P(X = x).$$

According to the definition, we have

$$\int_{-\infty}^{\infty} f(s)\,\mathrm{d}\,s = \lim_{x \to \infty} F(x) = 1$$

for continuous random variables, and

$$\sum_{x \in \operatorname{supp}(X)} p_x = 1$$

for discrete random variables.

***

Exercise 1.15. Two fair dice are rolled. Let X be the absolute value of the difference between the two numbers on the dice. Determine the probability function of X.

Exercise 1.16. A discrete random variable X with support $\{1, 2, 3, 4, 5\}$ has the following probability mass function:

$$p_1 = a^2,\ p_2 = 0.5a,\ p_3 = a,\ p_4 = 0.25a,\ p_5 = 0.5.$$

Determine the value of a.

Exercise 1.17. An actuary used the following model to model the number of claims filed by an individual under an automobile policy during a three-year period:

$$p_{n+1} = 0.2p_n, \quad n = 0, 1, 2, \ldots,$$

where p_n denotes the probability that a policyholder files n claims during the period. Calculate the probability that a policyholder files more than one claim during the period.

Exercise 1.18. The loss of an automobile claim has the following cdf:

$$F(x) = \begin{cases} \dfrac{3}{4}\left(\dfrac{x}{V}\right)^3, & \text{if } 0 \le x < V, \\ 1 - \dfrac{1}{10}\exp\left(-\dfrac{x - V}{V}\right), & \text{if } x \ge V, \end{cases}$$

where V is the value of the insured's automobile. Calculate the probability that the loss on a randomly selected claim is greater than the value of the automobile.

Exercise 1.19. Losses under a traveler's insurance policy are uniformly distributed on the interval $[0, 5]$. Under the policy, the insurer reimburses a policyholder for a loss up to a maximum of 4. Let X be the benefit that the insurer pays a policyholder who experiences exactly one loss. Determine the cumulative distribution function of X.

Exercise 1.20. Let X be a discrete random variable with probability function $p(x) = \frac{2}{3^x}$ for $x = 1, 2, 3, \ldots$. Calculate the probability that X is odd.

Exercise 1.21. Let X be a random variable with support $[2, \infty)$. Suppose that $f_X(x) = \frac{A}{x} S_X(x)$ and $F_X(5) = 0.84$. Determine A.

***

1.4 Transformations of Random Variables

Transformation is a technique that can be used to create new distributions. Common transformations include linear transformations, power transformations, and exponentiation. In this section, we introduce these common transformation and how to find the distribution of a transformation of a random variable.

Linear transformations are the simplest transformations and naturally occur when there is a change in measure units or reference points. Definition 1.12 gives the definition of linear transformations. Theorem 1.1 tells one how to find the distribution of the linear transformation of a random variable. The proof of Theorem 1.1 shows the standard way to find the distribution of a transformed random variable.

Definition 1.12 (Linear transformation)**.** Let X be a random variable. Let μ and $\sigma \neq 0$ be real numbers. Then the linear transformation of X is given by $\mu + \sigma X$, where μ is called the location parameter and σ is called the scale parameter.

Linear transformations are also called affine transformations.

Theorem 1.1 (Linear transformation)**.** *Let $Y = \mu + \sigma X$ be a linear transformation of a continuous random variable X, where μ and $\sigma \neq 0$ are real numbers. Then the cdf and the pdf of Y are given as follows:*

$$F_Y(y) = \begin{cases} F_X\left(\frac{y-\mu}{\sigma}\right), & \text{if } \sigma > 0, \\ 1 - F_X\left(\frac{y-\mu}{\sigma}\right), & \text{if } \sigma < 0, \end{cases}$$

$$f_Y(y) = \begin{cases} \dfrac{1}{\sigma} f_X\left(\dfrac{y-\mu}{\sigma}\right), & \text{if } \sigma > 0, \\ -\dfrac{1}{\sigma} f_X\left(\dfrac{y-\mu}{\sigma}\right), & \text{if } \sigma < 0, \end{cases}$$

where $F_X(\cdot)$ and $f_X(\cdot)$ are the cdf and the pdf of X, respectively.

Proof. Let us first consider the case when $\sigma > 0$. By the definition of the cdf and the assumption that $\sigma > 0$, we have

$$F_Y(y) = P(Y \le y) = P(\mu + \sigma X \le y) = P\left(X \le \frac{y-\mu}{\sigma}\right) = F_X\left(\frac{y-\mu}{\sigma}\right).$$

By the chain rule, we have

$$f_Y(y) = \frac{\mathrm{d}\, F_Y(y)}{\mathrm{d}\, y} = \frac{\mathrm{d}\, F_X\left(\frac{y-\mu}{\sigma}\right)}{\mathrm{d}\, y} = \frac{1}{\sigma} f_X\left(\frac{y-\mu}{\sigma}\right).$$

Now let us consider the case when $\sigma < 0$. In this case, we have

$$\begin{aligned} F_Y(y) &= P(Y \le y) = P(\mu + \sigma X \le y) = P\left(X \ge \frac{y-\mu}{\sigma}\right) \\ &= 1 - P\left(X < \frac{y-\mu}{\sigma}\right) = 1 - F_X\left(\frac{y-\mu}{\sigma}\right) + P\left(X = \frac{y-\mu}{\sigma}\right) \end{aligned}$$

Since X is continuous, it has zero point mass at any point, i.e., $P(X = x) = 0$ for all real number x. Hence

$$F_Y(y) = 1 - F_X\left(\frac{y-\mu}{\sigma}\right).$$

Taking the derivative, we get the following pdf:

$$f_Y(y) = -\frac{1}{\sigma} f_X\left(\frac{y-\mu}{\sigma}\right).$$

This completes the proof. □

Example 1.6 illustrates an application of Theorem 1.1 to find the pdf of a transformed random variable.

Example 1.6. Let $X \sim \mathrm{N}(0, 1)$, i.e., X is a standard normal random variable. Let μ and $\sigma > 0$ be real numbers. Find the pdf of $Y = \mu + \sigma X$.

Solution. By Theorem 1.1, we have

$$f_Y(y) = \frac{1}{\sigma} f_X\left(\frac{y-\mu}{\sigma}\right) = \frac{1}{\sqrt{2\pi}\sigma} \exp\left(\frac{-(y-\mu)^2}{2\sigma^2}\right).$$

We see that $Y \sim N(\mu, \sigma^2)$, i.e., Y is a normal random variable with mean μ and variance σ^2. □

Power transformations of a random variable involve raising the random variable to a power. Definition 1.13 gives the definition of power transformations. Theorem 1.2 gives the formulas for the cdf and the pdf of power-transformed random variables.

Definition 1.13 (Power transformation). Let X be a continuous random variable with positive support and $\tau \neq 0$. Then the power transformation of X is given by $X^{\frac{1}{\tau}}$.

When $\tau > 0$ ($\tau \neq 1$), $\tau = -1$, and $\tau < 0$ ($\tau \neq -1$), the resulting distribution is called transformed, inverse, and inverse transformed, respectively.

Theorem 1.2 (Power transformation). *Let X be a positive continuous random variable with cdf $F_X(x)$ and pdf $f_X(x)$. Let $Y = X^{\frac{1}{\tau}}$ be a power transformation of X. Then the cdf and the pdf of Y are given as follows:*

$$F_Y(y) = \begin{cases} F_X(y^\tau), & \text{if } \tau > 0, \\ 1 - F_X(y^\tau), & \text{if } \tau < 0, \end{cases}$$

$$f_Y(y) = \begin{cases} \tau y^{\tau-1} f_X(y^\tau), & \text{if } \tau > 0, \\ -\tau y^{\tau-1} f_X(y^\tau), & \text{if } \tau < 0. \end{cases}$$

Proof. First, let us consider the case when $\tau > 0$. By definition, we have for $y > 0$,

$$F_Y(y) = P(Y \leq y) = P\left(X^{\frac{1}{\tau}} \leq y\right) = P(X \leq y^\tau) = F_X(y^\tau).$$

The pdf can be obtained by taking the derivative:

$$f_Y(y) = \frac{\mathrm{d}\, F_Y(y)}{\mathrm{d}\, y} = f_X(y^\tau)\, \tau y^{\tau-1} = \tau y^{\tau-1} f_X(y^\tau).$$

Now let us consider the case when $\tau < 0$. The process is similar. However, when τ is negative, the inequality $X^{\frac{1}{\tau}} \leq y$ is equivalent to the inequality $X \geq y^\tau$. To see this, we can use the natural log function, which is an increasing function:

$$\ln \frac{X}{y^\tau} = \tau \left(\ln\left(X^{\frac{1}{\tau}}\right) - \ln y\right) \geq 0 = \ln 1.$$

Since X is continuous, the point mass of X is zero. Hence

$$\begin{aligned} F_Y(y) &= P(Y \leq y) = P\left(X^{\frac{1}{\tau}} \leq y\right) = P(X \geq y^\tau) \\ &= 1 - P(X < y^\tau) = 1 - P(X \leq y^\tau) + P(X = y^\tau) \\ &= 1 - F_X(y^\tau) + 0 = 1 - F_X(y^\tau). \end{aligned}$$

The pdf of X can obtained similarly. □

Note that Theorem 1.2 may not be correct when X can take negative values (see Exercise 1.23). In such cases, the cdf and the pdf of the power-transformed random variables cannot be obtained by Theorem 1.2.

Definition 1.14 (Exponentiation). Let X be a continuous random variable. Then the exponentiation transformation of X is given by

$$e^X.$$

Theorem 1.3. *Let X be a continuous random variable with cdf $F_X(x)$ and pdf $f_X(x)$. Let $Y = e^X$ be the exponentiation transformation of X. Then the cdf and the pdf of Y are given by:*

$$F_Y(y) = F_X(\ln y), \quad y > 0,$$

$$f_Y(y) = \frac{1}{y} f_X(\ln y), \quad y > 0.$$

Proof. Let $y > 0$. Then the cdf of Y is

$$F_Y(y) = P(Y \leq y) = P(e^X \leq y) = P(X \leq \ln y) = F_X(\ln y).$$

The pdf of Y can be obtained by

$$f_Y(y) = \frac{\mathrm{d}\, F_Y(y)}{\mathrm{d}\, y} = \frac{1}{y} f_X(\ln y).$$

□

***

Exercise 1.22. Let X be a random variable following the exponential distribution with parameter θ, i.e., $F_X(x) = 1 - \exp(-\frac{x}{\theta})$. Find the pdf of X^2.

Exercise 1.23. Let X be a standard normal random variables, i.e., $X \sim N(0, 1)$. Find the pdf of $Y = X^2$.

Exercise 1.24. Let X be a standard normal random variable. Find the pdf of $Y = |X|$.

Exercise 1.25. Let U be a uniform random variable on $[0, 1]$. Let X be a random variable with the following pdf:

$$f_X(x) = 3x^2, \quad 0 \leq x \leq 1.$$

Find the transformation g such that $g(U)$ and X have the same distribution.

***

1.5 Expectation

Expectation of a random variable is one of the most important concepts in probability theory. For a discrete random variable, its expectation is the weighted average of possible values that the random variable can take. For a continuous random variable, its expectation is defined by integration. Definition 1.15 gives the definition of expectation of a function of a random variable.

Definition 1.15 (Expectation)**.** Let X be a random variable and $g : \mathbb{R} \to \mathbb{R}$ be a function. The expectation of $g(X)$ is defined as follows:

$$E\left[g(X)\right] = \begin{cases} \int_{-\infty}^{\infty} g(x) f(x) \, \mathrm{d}\, x, & \text{if } X \text{ is continuous,} \\ \sum_{x \in \mathrm{supp}(X)} g(x) p(x), & \text{if } X \text{ is discrete,} \end{cases}$$

where $\mathrm{supp}(X)$ is the support of X.

Definition 1.16 (Raw moment)**.** The kth raw moment of a random variable X is defined as the expected value of X^k and is denoted by μ'_k or $E[X^k]$. It is calculated as follows:

$$\mu'_k = E\left[X^k\right] = \begin{cases} \int_{-\infty}^{\infty} x^k f(x) \, \mathrm{d}\, x, & \text{if } X \text{ is continuous,} \\ \sum_{x \in \mathrm{supp}(X)} x^k p(x), & \text{if } X \text{ is discrete,} \end{cases}$$

where $\mathrm{supp}(X)$ is the support of X. The first raw moment is called the mean of the random variable and is denoted by μ or $E[X]$.

For some random variables, their raw moments can be infinity. In such cases, we say that those raw moments do not exist.

Example 1.7. Let U be a uniform random variable on $[0, 1]$. Calculate the kth raw moment of U.

Solution. The random variable U has the following pdf: $f(x) = 1$ for $x \in (0, 1)$. Hence

$$\mu'_k = E\left[X^k\right] = \int_0^1 x^k \, \mathrm{d}\, x = \frac{1}{k+1} x^{k+1} \Big|_0^1 = \frac{1}{k+1}.$$

By setting $k = 1$, we get the mean of U as

$$E[X] = \frac{1}{2}.$$

□

Definition 1.17 (Central moments)**.** The kth central moment of a random variable X is defined as

$$\mu_k = E\left[(X-\mu)^k\right],$$

where μ is the mean of X. The second central moment is called the variance of the random variable and is denoted by σ^2 or $\text{Var}(X)$. The square root of the variance is called the standard deviation.

Central moments can be obtained from raw moments. Their relationship is summarized in Theorem 1.4. For example, when $k = 2$, we have

$$\begin{aligned}\mu_2 &= \mu_0'(-\mu)^2 + 2\mu_1'(-\mu) + \mu_2'(-\mu)^0 = \mu^2 - 2\mu^2 + \mu_2' \\ &= \mu_2' - \mu^2 = E\left[X^2\right] - E[X]^2.\end{aligned}$$

In the above derivation, we used $\mu_0' = E[X^0] = 1$ and $\mu_1' = E[X] = \mu$.

Theorem 1.4. *The kth central moment of a random variable X can be expressed by the raw moments:*

$$\mu_k = \sum_{i=0}^{k} \binom{k}{i} \mu_i'(-\mu)^{k-i},$$

where μ_i' is the ith raw moment of X, i.e., $\mu_i' = E\left[X^i\right]$.

Proof. By Definition 1.17 and the binomial formula, we have

$$\begin{aligned}\mu_k &= E\left[\sum_{i=0}^{k}\binom{k}{i} X^i(-\mu)^{k-i}\right] = \sum_{i=0}^{k}\binom{k}{i}(-\mu)^{k-i}E\left[X^i\right] \\ &= \sum_{i=0}^{k}\binom{k}{i}\mu_i'(-\mu)^{k-i}.\end{aligned}$$

□

Definition 1.18 (Coefficient of variation)**.** The coefficient of variation of a random variable is defined to be the ratio of its standard deviation to its mean, i.e.,

$$\text{CV} = \frac{\sigma}{\mu}.$$

The coefficient of variation shows the relative variability of a variable. When the mean is zero, the coefficient of variation is not defined. In addition, it is only meaningful when the mean is a valid measure of central tendency.

Definition 1.19 (Measures of shape)**.** The skewness of a random variable is defined to be the ratio of its third central moment to the cube of its standard deviation, i.e.,

$$\gamma_1 = \frac{\mu_3}{\sigma^3}.$$

The kurtosis is defined to be the ratio of its fourth central moment to the fourth power of its standard deviation, i.e.,

$$\gamma_2 = \frac{\mu_4}{\sigma^4}.$$

The skewness is a measure of asymmetry. A symmetric distribution has a skewness of zero, while a positive skewness indicates that probabilities to the right tend to be assigned to values further from the mean than those to the left.

The kurtosis measures the flatness of the distribution relative to a normal distribution (which has a kurtosis of 3). Kurtosis values above 3 indicate that (keeping the standard deviation constant), relative to a normal distribution, more probability tends to be at points away from the mean than at points near the mean.

Example 1.8. A random variable X has the following pdf:

$$f(x) = \begin{cases} \dfrac{x}{\theta}, & \text{if } 0 \le x < \theta, \\ \dfrac{-x+2}{2-\theta}, & \text{if } \theta \le x \le 2, \end{cases}$$

where $\theta \in (0, 2)$ is a parameter. Calculate the skewness of X.

Solution. To calculate the skewness of X, let us first calculate the first three raw moments. The first raw moment is

$$\begin{aligned} \mu = \mu_1' = \int_0^2 x f(x)\,\mathrm{d}x = \int_0^\theta \frac{x^2}{\theta}\,\mathrm{d}x + \int_\theta^2 \frac{-x^2+2x}{2-\theta}\,\mathrm{d}x \\ = \frac{\theta^2}{3} + \frac{\theta^3 - 3\theta^2 + 4}{3(2-\theta)} = \frac{\theta^2}{3} - \frac{\theta^2 - \theta - 2}{3} \\ = \frac{\theta + 2}{3}. \end{aligned}$$

The second raw moment is

$$\begin{aligned}\mu_2' = \int_0^2 x^2 f(x)\,\mathrm{d}\,x = & \int_0^\theta \frac{x^3}{\theta}\,\mathrm{d}\,x + \int_\theta^2 \frac{-x^3+2x^2}{2-\theta}\,\mathrm{d}\,x \\ = & \frac{\theta^3}{4} + \frac{3\theta^4 - 8\theta^3 + 16}{12(2-\theta)} = \frac{-\theta^3+8}{6(2-\theta)} \\ = & \frac{\theta^2+2\theta+4}{6}.\end{aligned}$$

The third raw moment is

$$\begin{aligned}\mu_3' = \int_0^2 x^3 f(x)\,\mathrm{d}\,x = & \int_0^\theta \frac{x^4}{\theta}\,\mathrm{d}\,x + \int_\theta^2 \frac{-x^4+2x^3}{2-\theta}\,\mathrm{d}\,x \\ = & \frac{\theta^4}{5} + \frac{2\theta^5 - 5\theta^4 + 16}{10(2-\theta)} = \frac{-\theta^4+16}{10(2-\theta)} \\ = & \frac{(\theta^2+4)(\theta+2)}{10}.\end{aligned}$$

From the raw moments, we can get the variance

$$\mathrm{Var}(X) = \mu_2' - \mu^2 = \frac{\theta^2+2\theta+4}{6} - \left(\frac{\theta+2}{3}\right)^2 = \frac{\theta^2-2\theta+4}{18}.$$

By Theorem 1.4, we can get the third central moment

$$\begin{aligned}\mu_3 = & \mu_3' - 3\mu_2'\mu + 3\mu_1'\mu^2 - 3\mu^3 = \mu_3' - 3\mu_2'\mu + 2\mu^3 \\ = & \frac{(\theta^2+4)(\theta+2)}{10} - 3\cdot\frac{\theta^2+2\theta+4}{6}\cdot\frac{\theta+2}{3} + 2\left(\frac{\theta+2}{3}\right)^3 \\ = & \frac{(\theta+2)(\theta-1)(\theta-4)}{135}.\end{aligned}$$

Hence the skewness of X is

$$\gamma_1 = \frac{\mu_3}{\sqrt{\mu_2}} = \frac{\sqrt{2}(\theta+2)(\theta-1)(\theta-4)}{45\sqrt{\theta^2-2\theta+4}}.$$

When $\theta = 1$, the skewness of X is zero. When $\theta < 1$, the skewness of X is positive. When $\theta > 1$, X has negative skewness. Figure 1.2 shows the probability density functions of X when $\theta = 0.5$ and $\theta = 1.5$. We can see that when skewness is positive, the probability density function is skewed to the right side. When skewness is negative, the probability density function is skewed to the left side. □

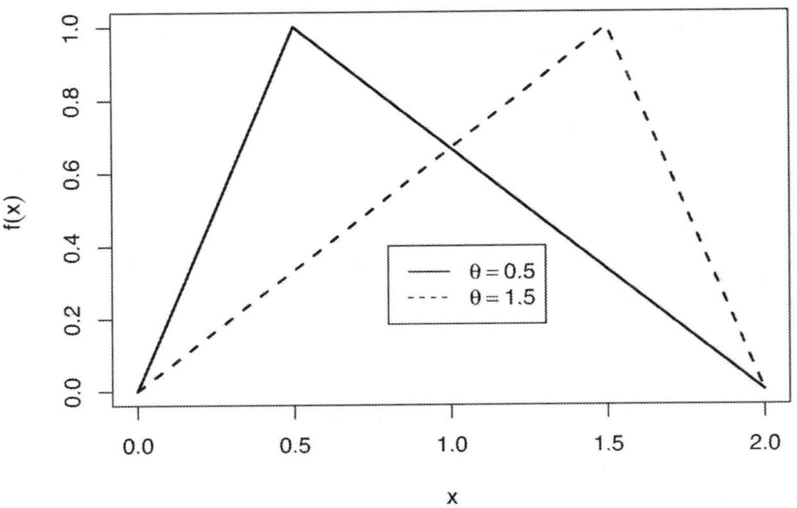

FIGURE 1.2
Probability density functions of the random variable given in Example 1.8.

Definition 1.20 (Percentile)**.** Let $p \in [0, 1]$. The 100pth percentile, also referred to as the p quantile, of a random variable X is any value π_p that satisfies the following condition:

$$F(\pi_p-) \le p \le F(\pi_p),$$

where $F(\cdot)$ is the cdf of X and $F(\pi_p-) = \lim_{x \uparrow \pi_p} F(x)$. The 50th percentile, $\pi_{0.5}$, is called the median.

Definition 1.21 (Moment-generating function and probability-generating function)**.** The moment-generating function (mgf) of a random variable X is defined as

$$M_X(z) = E\left[e^{zX}\right].$$

The probability-generating function (pgf) of X is defined as

$$P_X(z) = E\left[z^X\right].$$

Theorem 1.5 (Moments). *The mgf can be used to calculate raw moments as follows:*

$$\mu'_k = E\left[X^k\right] = M_X^{(k)}(0) = \left.\frac{\mathrm{d}^k M_X(z)}{\mathrm{d}\, z^k}\right|_{z=0}, \quad k = 0, 1, 2, \ldots.$$

Theorem 1.6 (Sum of random variables). *Let X_1, X_2, ... be a sequence of random variables. Let*

$$S_k = X_1 + X_2 + \cdots + X_k, \quad k \geq 1.$$

Then

(a) $E[S_k] = E[X_1] + E[X_2] + \cdots + E[X_k]$.

(b) If X_1, X_2, ..., X_k are independent, then

$$\mathrm{Var}(S_k) = \sum_{i=1}^{k} \mathrm{Var}(X_i).$$

(c) If X_1, X_2, ..., X_k are independent, then

$$M_{S_k}(z) = \prod_{i=1}^{k} M_{X_i}(z), \quad P_{S_k}(z) = \prod_{i=1}^{k} P_{X_i}(z).$$

***

Exercise 1.26. A random variable X has the following probability density function:

$$f(x) = \begin{cases} \dfrac{|x|}{10}, & \text{if } -2 \leq x \leq 4, \\ 0, & \text{otherwise.} \end{cases}$$

Calculate $E[X]$.

Exercise 1.27. A health insurance policy pays \$100 per day for the first three days of hospitalization and \$50 per day for the remaining days of hospitalization. The number of days of hospitalization, X, has the following probability function:

$$p_k = \frac{6-k}{15}, \quad k = 1, 2, 3, 4, 5.$$

Calculate the expected payment for hospitalization under this insurance policy.

Exercise 1.28. A device is placed in a remote region to continuously measure and record seismic activities. The time to failure of this device, T, is exponentially distributed with a mean of three years:

$$f(t) = \frac{1}{3}e^{-t/3}, \quad t > 0.$$

Since the device will not be monitored during its first two years of service, the time to discovery of its failure is $X = \max(T, 2)$. Calculate $E[X]$.

Exercise 1.29. Two random variables X and Y are independent and have the same mean. The coefficients of variation of X and Y are 3 and 4, respectively. Calculate the coefficient of variation of $\dfrac{X+Y}{2}$.

Exercise 1.30. Losses have a Pareto distribution with parameters α and θ. The 20th percentile is $\theta - k$. The 90th percentile is $3\theta - 2k$. Determine the value of α.

Exercise 1.31. The loss, X, incurred by a policyholder follows a normal distribution with a mean of 20,000 and a standard deviation of 4,500. The amount paid by the insurance policy is $Y = \max(0, X - 15,000)$. Calculate the 95th percentile of Y. (Hint: use the normal distribution table given in Table C.1.)

Exercise 1.32. At a large university, students can register for a semester's courses during a 13-day period. Let X be the number of days that elapsed before a randomly selected student registers. The pdf of X, $f(t)$, is continuous, symmetric about $t = 6.5$, and proportional to $1/(t+1)$ between days 0 and 6.5. A student registers at the 60th percentile of this distribution. Calculate the number of elapsed days in the period for this student.

Exercise 1.33. Let $a \in (0, 80)$ be a constant. A random variable X has the following pdf:

$$f_X(x) = \frac{1}{100 - a}, \quad x \in [a, 100].$$

Another random variable Y has the following pdf:

$$f_Y(y) = \frac{1}{100 - 1.25a}, \quad y \in [1.25a, 100].$$

Suppose that $E\left[X^2\right] = \dfrac{19600}{3}$. Calculate the 80th percentile of Y.

---***

1.6 Joint Distributions

Joint distributions involve two or more random variables. The joint cumulative distribution function of two random variables is defined in Definition 1.22.

Definition 1.22 (Joint cumulative distribution function)**.** Let X and Y be two random variables on a probability space $(\Omega, \mathscr{F}, P)$. Then the joint cumulative distribution function of X and Y is defined as

$$F(x, y) = P(X \leq x, Y \leq y), \quad x, y \in (-\infty, \infty).$$

Here $P(X \leq x, Y \leq y)$ is calculated as

$$P(X \leq x, Y \leq y) = P(\{\omega : X(\omega) \leq x \text{ and } Y(\omega) \leq y\}).$$

Definition 1.23 (Joint probability density function)**.** Let X and Y be two continuous random variables with the joint cumulative distribution function $F(x, y)$. Then their joint probability density function is defined as

$$f(x, y) = \frac{\partial^2 F(x, y)}{\partial x \partial y}.$$

Definition 1.24 (Joint probability mass function)**.** Let X and Y be two discrete random variables with the joint cumulative distribution function $F(x, y)$. Then their joint probability mass function is defined as

$$p(x, y) = P(X = x, Y = y).$$

For simplicity, we consider only two random variables in Definitions 1.22, 1.23, and 1.24. We can extend these definitions to more than two random variables straightforwardly. For example, the joint cdf for n random variables X_1, X_2, ..., X_n is defined as

$$F(x_1, x_2, \ldots, x_n) = P(X_1 \leq x_1, X_2 \leq x_2, \ldots, X_n \leq x_n).$$

Given the joint cumulative distribution function of two random variables, we can define the joint probability density function or the joint probability mass function. Definitions 1.23 and 1.24 give the definitions when both random variables have the same type. When one variable is continuous and another variable is discrete, we can also define joint density. Suppose that X is continuous and Y is discrete. For example, we can define the joint density as follows:

$$f(x, y) = f(x|y)P(Y = y) = P(Y = y)\frac{\partial F(x, y)}{\partial x},$$

where $f(x|y)$ is the probability density function of X given $Y = y$.

Given the joint probability density function or the joint probability mass function of two random variables, we can calculate the joint cumulative distribution function. For continuous random variables, we have

$$F(x,y) = \int_{-\infty}^{x} \int_{-\infty}^{y} f(s,t)\,\mathrm{d}s\,\mathrm{d}t.$$

For discrete random variables, we have

$$F(x,y) = \sum_{s\le x} \sum_{t\le y} p(s,t).$$

Example 1.9. In a year, the number of tornadoes, N, experienced by a homeowner has the following probability distribution:

$$p_N(0) = 0.8, \quad p_N(1) = 0.12, \quad p_N(2) = 0.05, \quad p_N(3) = 0.03.$$

A loss caused by a tornado, X, has the following distribution:

$$p_X(1) = 0.5, \quad p_X(2) = 0.5.$$

Let S be the total amount of losses experienced by the homeowner experiences in the year due to all the tornadoes. Let $F(n,s)$ be the joint cumulative distribution function of N and S. Suppose that the losses caused by tornadoes are independent. Calculate $F(2,3)$.

Solution. By Definition 1.22 and the property of probability measures, we have

$$\begin{aligned} F(2,3) =& P(N \le 2, S \le 3) = P(N=0, S\le 3) + P(N=1, S\le 3) \\ &+ P(N=2, S\le 3). \end{aligned}$$

When $N=0$, there will be no losses. Hence $P(N=0, S\le 3) = P(N=0) = 0.8$. When $N=1$, there is one loss and the total amount can be 1 or 2. Hence $P(N=1, S\le 3) = P(N=1) = 0.12$. When $N=2$, there are two losses and the total amount can be 2, 3, or 4. In this case,

$$\begin{aligned} P(N=2, S\le 3) =& P(N=2, S=2) + P(N=2, S=3) \\ =& P(N=2, X_1=1, X_2=1) + P(N=2, X_1=1, X_2=2) \\ &+ P(N=2, X_1=2, X_2=1) \\ =& 0.05 \times 0.5^2 + 0.05 \times 0.5^2 + 0.05 \times 0.5^2 \\ =& 0.0375. \end{aligned}$$

Hence

$$F(2,3) = 0.8 + 0.12 + 0.0375 = 0.9575.$$

□

Given the joint cdf, $F(x, y)$, of two random variables X and Y, we can get the marginal distributions as follows:

$$F_X(x) = P(X \le x) = P(X \le x, Y \le \infty) = F(x, \infty),$$
$$F_Y(y) = F(\infty, y).$$

For continuous random variables, the marginal distributions can also be obtained from the joint pdf:

$$F_X(x) = \int_{-\infty}^{\infty} f(x, y)\,\mathrm{d}\,y, \quad F_Y(y) = \int_{-\infty}^{\infty} f(x, y)\,\mathrm{d}\,x.$$

For discrete random variables, the marginal distributions are calculated as:

$$p_X(x) = \sum_{y \in \text{supp}(Y)} p(x, y), \quad p_Y(y) = \sum_{x \in \text{supp}(X)} p(x, y).$$

Example 1.10. Let X and Y be two discrete random variables with the following joint probability mass function:

$$p(x, y) = \begin{cases} \dfrac{-2x - 4y + xy + 8}{18}, & \text{if } x = 1, 2, 3,\ y = 0, 1 \\ 0, & \text{otherwise.} \end{cases}$$

Determine the marginal distributions of X and Y.

Solution. The marginal probability mass function of X is determined as follows:

$$\begin{aligned} p_X(x) =& p(x, 0) + p(x, 1) = \frac{-2x + 8}{18} + \frac{-2x - 4 + x + 8}{18} \\ =& \frac{-3x + 12}{18} = \frac{-x + 4}{6}, \quad x = 1, 2, 3. \end{aligned}$$

Similarly, the marginal probability mass function of Y is

$$\begin{aligned} P_Y(y) =& p(1, y) + p(2, y) + p(3, y) \\ =& \frac{-2 - 4y + y + 8}{18} + \frac{-4 - 4y + 2y + 8}{18} + \frac{-6 - 4y + 3y + 8}{18} \\ =& \frac{-6y + 12}{18} = \frac{-y + 2}{3}, \quad y = 0, 1. \end{aligned}$$

□

Definition 1.25 (Independence)**.** Two random variables X and Y are said to be independent if

$$P(\{X \in A, Y \in B\}) = P(X \in A)P(Y \in B)$$

for any two sets of real numbers A and B. Random variables that are not independent are said to be dependent.

Random variables X_1, X_2, ..., X_n are said to be independent if

$$P(\{X_1 \in A_1, X_2 \in A_2, \ldots, X_n \in A_n\}) = \prod_{i=1}^{n} P(X_i \in A_i)$$

for any sets of real numbers A_1, A_2, ..., A_n.

Definition 1.26 (Independent and identically distributed random variables)**.** Random variables X_1, X_2, ..., X_n are said to be independent and identically distributed (i.i.d.) if the n random variables are independent and have the same distribution.

Theorem 1.7. *If X and Y are independent random variables, then for any functions $g(x)$ and $h(y)$,*

$$E[g(X)h(Y)] = E[g(X)]E[h(Y)].$$

Theorem 1.8. *Let X and Y be two random variables with the joint cdf $F(x, y)$. Let $F_X(x)$ and $F_Y(y)$ be the cdfs of X and Y, respectively. Then X and Y are independent if and only if*

$$F(a, b) = F_X(a)F_Y(b)$$

for all real numbers a and b.

When X and Y are continuous, then X and Y are independent if and only if

$$f(a, b) = f_X(a)f_Y(b)$$

for all a and b, where $f(x, y)$ is the joint pdf of X and Y, $f_X(x)$ is the pdf of X, and $f_Y(y)$ is the pdf of Y.

When X and Y are discrete, then X and Y are independent if and only if

$$p(a, b) = p_X(a)p_Y(b)$$

for all a and b, where $p(x, y)$ is the joint probability mass function of X and Y, $p_X(x)$ is the probability mass function of X, and $p_Y(y)$ is the probability mass function of Y.

Example 1.11. Let X and Y be two random variables with the following joint pdf:

$$f(x, y) = 6e^{-2x}e^{-3y}, \quad 0 < x, y < \infty.$$

Determine whether X and Y are independent.

Solution. Let us calculate the marginal pdfs of X and Y. The marginal pdf of X is

$$f_X(x) = \int_0^\infty f(x,y)\,\mathrm{d}\,y = \int_0^\infty 6e^{-2x}e^{-3y}\,\mathrm{d}\,y = 2e^{-2x}.$$

Similarly, the marginal pdf of Y is

$$f_Y(y) = \int_0^\infty f(x,y)\,\mathrm{d}\,x = \int_0^\infty 6e^{-2x}e^{-3y}\,\mathrm{d}\,y = 3e^{-2y}.$$

Since

$$f_X(x)f_Y(y) = 6e^{-2x}e^{-3y} = f(x,y),$$

it follows from Theorem 1.8 that X and Y are independent. □

Example 1.12. Let X and Y be two random variables with the following joint pdf:

$$f(x,y) = 24xy, \quad 0 < x, y < 1,\ 0 < x+y < 1.$$

Determine whether X and Y are independent.

Solution. To determine whether X and Y are independent, we can check whether the joint pdf is the product of the marginal pdfs. The marginal pdf of X is

$$f_X(x) = \int_0^{1-x} 24xy\,\mathrm{d}\,y = 12x(1-x)^2, \quad 0 < x < 1.$$

The marginal pdf of Y is

$$f_Y(y) = \int_0^{1-y} f(x,y)\,\mathrm{d}\,x = 12y(1-y)^2, \quad 0 < y < 1.$$

Since

$$f_X(x)f_Y(y) = 144xy(1-x)^2(1-y)^2 \neq f(x,y),$$

it follows from Theorem 1.8 that X and Y are not independent. □

Definition 1.27 (Covariance)**.** Let X and Y be two random variables. The covariance between X and Y is defined by

$$\mathrm{Cov}(X,Y) = E[(X - E[X])(Y - E[Y])].$$

Definition 1.28 (Correlation coefficient)**.** Let X and Y be two random variables. The correlation coefficient between X and Y is defined by

$$\mathrm{Corr}(X,Y) = \frac{\mathrm{Cov}(X,Y)}{\sqrt{\mathrm{Var}(X)\,\mathrm{Var}(Y)}}.$$

Theorem 1.9. *The covariance has the following properties:*

(a) $\operatorname{Cov}(X,Y) = E[XY] - E[X]E[Y]$.

(b) $\operatorname{Cov}(X,Y) = \operatorname{Cov}(Y,X)$.

(c) $\operatorname{Cov}(X,X) = \operatorname{Var}(X)$.

(d) $\operatorname{Cov}(aX,Y) = a\operatorname{Cov}(X,Y)$ *for all* a.

(e)

$$\operatorname{Cov}\left(\sum_{i=1}^{n} X_i, \sum_{j=1}^{m} Y_j\right) = \sum_{i=1}^{n}\sum_{j=1}^{m} \operatorname{Cov}(X_i, Y_j).$$

Definition 1.29 (Conditional distribution)**.** Let X and Y be continuous random variables with the joint pdf $f(x,y)$. Then the conditional pdf of X given $Y = y$ is defined, for all values of y such that $f_Y(y) > 0$, by

$$f_{X|Y}(x|y) = \frac{f(x,y)}{f_Y(y)},$$

where $f_Y(y)$ is the marginal pdf of Y. The conditional expectation of X given $Y = y$ is defined by

$$E[X|Y=y] = \int_{-\infty}^{\infty} x f_{X|Y}(x|y)\,\mathrm{d}\,x = \frac{1}{f_Y(y)}\int_{-\infty}^{\infty} x f(x,y)\,\mathrm{d}\,x.$$

Let X and Y be discrete random variables with the joint pf $p(x,y)$. Then the conditional pf of X given $Y = y$ is defined, for all values of y such that $p_Y(y) > 0$, by

$$p_{X|Y}(x|y) = \frac{p(x,y)}{p_Y(y)},$$

where $p_Y(y)$ is the marginal pf of Y. The conditional expectation of X given $Y = y$ is defined by

$$E[X|Y=y] = \sum_{x\in\operatorname{supp}(X)} x p_{X|Y}(x|y) = \frac{1}{p_Y(y)} \sum_{x\in\operatorname{supp}(X)} x p(x,y).$$

The symbol $E[X|Y]$ denotes the function of the random variable Y such that its value at y is $E[X|Y=y]$.

Theorem 1.10 (Conditional expectation formula). *Let X and Y be two random variables. Then*

$$E[X] = E[E[X|Y]].$$

The above formula is also called the tower property of expectation.

Proof. Note that $E[X|Y]$ is a function of Y. When Y is discrete, we have

$$\begin{aligned} E[E[X|Y]] &= \sum_{y \in \text{supp}(Y)} E[X|Y=y] p_Y(y) = \sum_{y \in \text{supp}(Y)} \sum_{x \in \text{supp}(X)} x p(x,y) \\ &= E[X]. \end{aligned}$$

When Y is continuous, we have

$$\begin{aligned} E[E[X|Y]] &= \int_{-\infty}^{\infty} E[X|Y=y] f_Y(y) \, \mathrm{d}\, y = \int_{-\infty}^{\infty} \int_{-\infty}^{\infty} x f(x,y) \, \mathrm{d}\, x \, \mathrm{d}\, y \\ &= E[X]. \end{aligned}$$

□

Example 1.13. Let X be a random variable that indicates the presence of a disease in a patient: $X = 1$ if the disease is present; $X = 0$ if the disease is not present. Let Y be the outcome of a diagnostic test of the disease: $Y = 1$ if the disease is tested to be present; $Y = 0$ if the disease is tested to be absent. The joint pf of X and Y, $p(x,y)$, is

$$p(0,0) = 0.8, \quad p(1,0) = 0.05, \quad p(0,1) = 0.025, \quad p(1,1) = 0.125.$$

Calculate $\text{Var}(Y|X=1)$.

Solution. Let us first determine the conditional pf of Y given $X = 1$. By Definition 1.29, we have

$$p_{Y|X}(0|1) = \frac{p(1,0)}{p_X(1)} = \frac{p(1,0)}{p(1,0)+p(1,1)} = \frac{0.05}{0.05+0.125} = \frac{2}{7}$$

and

$$p_{Y|X}(1|1) = \frac{p(1,1)}{p_X(1)} = \frac{p(1,1)}{p(1,0)+p(1,1)} = \frac{0.125}{0.05+0.125} = \frac{5}{7}.$$

By using the conditional pf, we have

$$E[Y|X=1] = 0 \times p_{Y|X}(0|1) + 1 \times p_{Y|X}(1,1) = \frac{5}{7}$$

and

$$E[Y^2|X=1] = 0^2 \times p_{Y|X}(0|1) + 1^2 \times p_{Y|X}(1,1) = \frac{5}{7}.$$

Hence the conditional variance is

$$\mathrm{Var}(Y|X=1) = E[Y^2|X=1] - E[Y|X=1]^2 = \frac{5}{7} - \left(\frac{5}{7}\right)^2 = \frac{10}{49}.$$

□

***

Exercise 1.34. Let X and Y be two random variables defined on a probability space $(\Omega, \mathscr{F}, P)$. Let $F(x, y)$ be the joint cdf of X and Y. Let $a_1 < a_2$ and $b_1 < b_2$. Show that

$$P(a_1 < X \le a_2, b_1 < Y \le b_2) = F(a_2, b_2) + F(a_1, b_1) - F(a_1, b_2) - F(a_2, b_1).$$

Exercise 1.35. Claim amounts are independent and identically distributed random variables that have the following pdf:

$$f(x) = \frac{10}{x^2}, \quad x > 10.$$

Calculate the probability that the largest of three randomly selected claims is less than 25.

Exercise 1.36. Let X and Y denote the number of tornadoes in two countries during a year. The joint pf of X and Y is given below:

		Y			
		0	1	2	3
	0	0.12	0.06	0.05	0.02
X	1	0.13	0.15	0.12	0.03
	2	0.05	0.15	0.10	0.02

Calculate $\mathrm{Var}(Y|X=0)$.

Exercise 1.37. Let X and Y be the number of claims submitted to a life insurance in April and May, respectively. The joint pf of X and Y is

$$p(x, y) = \frac{3e^{-x}(1 - e^{-x})^{y-1}}{4^x}, \quad x, y = 1, 2, 3, \ldots.$$

Calculate $E[Y|X=2]$.

Exercise 1.38. Let X and Y be two discrete random variables such that $\mathrm{supp}(X) = \{0, 1\}$ and $\mathrm{supp}(Y) = \{0, 1, 2\}$. Let $p(x, y)$ be the joint pf of X and Y with the following properties:

(a) $p(1, 2) = 3p(1, 1)$.

(b) $p(1, 1)$ maximizes the variance of XY.

Calculate the probability that $X = 0$ or $Y = 0$.

***

1.7 Compound Distributions

A compound distribution refers to a probability distribution whose parameters are also random variables. Compound distributions are also called mixture distributions or contagious distributions.

The cdf and the pdf of a compound distribution are usually denoted by $F_{X|\Lambda}(x|\lambda)$ and $f_{X|\Lambda}(x|\lambda)$, respectively, where Λ is a parameter of X. The unconditional cdf and the unconditional pdf of X can be obtained by

$$F_X(x) = \int_{\text{supp}(\Lambda)} F_{X|\Lambda}(x|\lambda) f_\Lambda(\lambda) \, \mathrm{d}\, \lambda \tag{1.1}$$

and

$$f_X(x) = \int_{\text{supp}(\Lambda)} f_{X|\Lambda}(x|\lambda) f_\Lambda(\lambda) \, \mathrm{d}\, \lambda, \tag{1.2}$$

where $f_\Lambda(\lambda)$ is the pdf of Λ.

Theorem 1.11. *Let X be a compound random variable with a parameter Λ. Then the unconditional raw moments are given by*

$$E\left[X^k\right] = E\left[E\left[X^k|\Lambda\right]\right], \quad k = 1, 2, \ldots.$$

In particular,

$$\text{Var}(X) = E[\text{Var}(X|\Lambda)] + \text{Var}(E[X|\Lambda]). \tag{1.3}$$

Proof. Note that $E[X^k|\Lambda]$ is a function of Λ whose value at λ is $E[X^k|\Lambda = \lambda]$ (see Definition 1.29). We have

$$\begin{aligned}
E\left[E\left[X^k|\Lambda\right]\right] &= \int_{\text{supp}(\Lambda)} E\left[X^k|\Lambda = \lambda\right] f_\Lambda(\lambda) \, \mathrm{d}\, \lambda \\
&= \int_{\text{supp}(\Lambda)} \int_{\text{supp}(X)} x^k f_{X|\Lambda}(x|\lambda) \, \mathrm{d}\, x f_\Lambda(\lambda) \, \mathrm{d}\, \lambda \\
&= \int_{\text{supp}(X)} x^k \int_{\text{supp}(\Lambda)} f_{X|\Lambda}(x|\lambda) f_\Lambda(\lambda) \, \mathrm{d}\, \lambda \, \mathrm{d}\, x \\
&= \int_{\text{supp}(X)} x^k f_X(x) \, \mathrm{d}\, x = E[X^k].
\end{aligned}$$

The equation for the variance can be shown as follows:

$$\begin{aligned}
&E[\text{Var}(X|\Lambda)] + \text{Var}(E[X|\Lambda]) \\
=&E\left(E[X^2|\Lambda] - E[X|\lambda]^2\right) + E[E[X|\Lambda]^2] - (E[E[X|\Lambda]])^2 \\
=&E[X^2] - E[X]^2 = \text{Var}(X).
\end{aligned}$$

□

Example 1.14. Let X be an exponential random variable with parameter Θ, i.e.,

$$f_{X|\Theta}(x|\theta) = \frac{1}{\theta}\exp\left(-\frac{x}{\theta}\right), \quad x > 0.$$

Suppose that Θ is uniformly distributed on $(0, 1)$. Computer $\mathrm{Var}(X)$.

Solution. Let us first calculate the conditional mean and the conditional variance of X. The conditional mean is

$$E[X|\Theta = \theta] = \int_0^\infty \frac{x}{\theta} e^{-\frac{x}{\theta}}\,\mathrm{d}\,x = \theta.$$

Hence $E[X|\Theta] = \Theta$. The conditional second raw moment of X is

$$E[X^2|\Theta = \theta] = \int_0^\infty \frac{x^2}{\theta} e^{-\frac{x}{\theta}}\,\mathrm{d}\,x = 2\theta^2.$$

Hence $E[X^2|\Theta] = 2\Theta^2$. We have $\mathrm{Var}(X|\Lambda) = \Theta^2$. Now by Theorem 1.11, we have

$$\mathrm{Var}(X) = E[\Theta^2] + \mathrm{Var}(\Theta) = 2E[\Theta^2] - E[\Theta]^2.$$

Since Θ is uniform on $(0, 1)$, we have

$$E[\Theta] = \int_0^1 \theta\,\mathrm{d}\,\theta = \frac{1}{2}, \quad E[\Theta^2] = \int_0^1 \theta^2\,\mathrm{d}\,\theta = \frac{1}{3}.$$

Therefore,

$$\mathrm{Var}(X) = \frac{2}{3} - \frac{1}{4} = \frac{5}{12}.$$

□

***

Exercise 1.39. The random variable X is the number of dental claims in a year and has the following pdf:

$$f_{X|\Theta}(x|\theta) = \frac{1}{\theta}e^{-x/\theta}, \quad x > 0.$$

The parameter Θ is distributed uniformly between 1 and 3. Calculate $E(X)$ and $\mathrm{Var}(X)$.

Exercise 1.40. Let N be a compound Poisson random variable with the following pf:

$$p_{N|\Lambda}(k|\lambda) = \frac{\lambda^k e^{-\lambda}}{k!}, \quad k = 0, 1, 2, \ldots.$$

Let Λ have a uniform distribution on the interval (0,5). Determine the unconditional probability that $N > 2$.

Exercise 1.41. Let N be a compound Poisson random variable with the following pf:

$$p_{N|\Lambda}(k|\lambda) = \frac{\lambda^k e^{-\lambda}}{k!}, \quad k = 0, 1, 2, \ldots.$$

Let Λ have a gamma distribution withe the following pdf:

$$f_\Lambda(\lambda) = \frac{(\lambda/\theta)^\alpha e^{-\lambda/\theta}}{\lambda\Gamma(\alpha)}, \quad \lambda > 0,$$

where $\alpha > 0$ and $\theta > 0$. Find the unconditional pf of N.

***

1.8 Limit Theorems

Limit theorems are important theoretical results in probability theory. In this section, we present the concepts of convergence and some important limit theorems.

Definition 1.30 (Convergence in distribution)**.** A sequence of random variables $\{X_n\}_{n\geq 1}$ is said to converge in distribution (or converge weekly, or converge in law) to X if for all x,

$$\lim_{n\to\infty} F_n(x) = F(x),$$

where $F_n(x)$ is the cdf of X_n and $F(x)$ is the cdf of X.

Definition 1.31 (Convergence in probability)**.** A sequence of random variables $\{X_n\}_{n\geq 1}$ is said to converge in probability to X if for all $\epsilon > 0$,

$$\lim_{n\to\infty} P\left(|X_n - X| > \epsilon\right) = 0.$$

Definition 1.32 (Convergence almost surely)**.** A sequence of random variables $\{X_n\}_{n\geq 1}$ is said to converge almost surely (or almost everywhere or strongly or with probability 1) to X if

$$P\left(\lim_{n\to\infty} X_n = X\right) = 1.$$

Definitions 1.30, 1.31, and 1.32 define three forms of convergence that are common in probability theory. Convergence in distribution is the weakest form. Convergence almost surely is the strongest form. In fact, convergence almost surely implies convergence in probability, which in turn implies convergence in distribution.

Theorem 1.12 (Weak law of large numbers). *Let* $\{X_n\}_{n\geq 1}$ *be a sequence of i.i.d. random variables that have finite mean* $E[X_i] = \mu$. *Then the running mean* $\frac{X_1 + X_2 + \cdots + X_n}{n}$ *converges to* μ *in probability, i.e.,*

$$\lim_{n\to\infty} P\left(\left|\frac{X_1 + X_2 + \cdots + X_n}{n} - \mu\right| > \epsilon\right) = 0, \quad \forall \epsilon > 0.$$

Theorem 1.13 (Strong law of large numbers). *Let* $\{X_n\}_{n\geq 1}$ *be a sequence of i.i.d. random variables that have finite mean* $E[X_i] = \mu$. *Then the running mean* $\frac{X_1 + X_2 + \cdots + X_n}{n}$ *converges almost surely to* μ, *i.e.,*

$$P\left(\lim_{n\to\infty} \frac{X_1 + X_2 + \cdots + X_n}{n} = \mu\right) = 1.$$

Theorems 1.12 and 1.13 are two versions of the law of large numbers. The weak law of large numbers says that the running average converges to the expected value in probability; while the strong law of large numbers says the convergence is almost everywhere.

Theorem 1.14 (Central limit theorem). *Let* $\{X_n\}_{n\geq 1}$ *be a sequence of i.i.d. random variables that have finite mean* $E[X_i] = \mu$ *and variance* $\mathrm{Var}(X_i) = \sigma^2$. *Then*

$$\frac{X_1 + X_2 + \cdots + X_n - n\mu}{\sigma\sqrt{n}}$$

converges in distribution to the standard normal distribution, i.e.,

$$\lim_{n\to\infty} P\left(\frac{X_1 + X_2 + \cdots + X_n - n\mu}{\sigma\sqrt{n}} \leq x\right) = \Phi(x), \quad \forall x \in (-\infty, \infty).$$

Here $\Phi(x)$ *is the cdf of the standard normal distribution (see Appendix C).*

The central limit theorem given in Theorem 1.14 is one of the most important theoretical results in probability theory. It states that the sum of a large number of i.i.d. random variables can be approximated by the normal distribution. For proofs of these limit theorems, readers are referred to [17].

Example 1.15. Let $\{X_n\}_{n\geq 1}$ be a sequence of i.i.d. random variables that follow the exponential distribution with parameter 1, i.e., $P(X_1 \leq x) = 1 - e^{-x}$. For $n \geq 1$, let $M_n = \max(X_1, X_2, \ldots, X_n)$. Show that $M_n - \ln n$ converges weekly.

Solution. Let us calculate the cdf of $M_n - \ln n$. This can be done as follows:

$$\begin{aligned}
P(M_n - \ln n \leq x) =& P\left(\max(X_1, X_2, \ldots, X_n) \leq \ln n + x\right) \\
=& P(X_1 \leq \ln n + x, X_2 \leq \ln n + x, \ldots, X_n \leq \ln n + x) \\
=& P(X_1 \leq \ln n + x)^n = (1 - e^{-x - \ln n})^n \\
=& \left(1 + \frac{-e^{-x}}{n}\right)^n.
\end{aligned}$$

In the above derivation, we used the assumption that X_1, X_2, ..., X_n are i.i.d. and follow the exponential distribution.

By Theorem A.6, we have

$$\lim_{n\to\infty} P(M_n - \ln n \leq x) = e^{-e^{-x}}, \quad x \in (-\infty, \infty).$$

The random variable $M_n - \ln n$ converges in distribution to a Gumbel distribution. □

***

Exercise 1.42. A company provides life insurance for each of its 1,000 employees. There is a 1.4% chance that any one employee will die next year, independent of all other employees. The death benefit is $50,000. The company plans to establish a fund to cover next year's death benefits with a probability of at least 99%. Calculate the smallest amount of money that the company must put into the fund. Round the answer to the nearest thousand.

Exercise 1.43. An insurance company has a portfolio of 10,000 automobile insurance policies. The yearly claims from the policies are i.i.d. with a common mean of $240 and a common standard deviation of $800. Approximate the probability that the total yearly claim from the portfolio exceeds $2.5 million.

Exercise 1.44. An insurance company has a portfolio of 1,250 health insurance policies. The number of claims filed by a policyholder during a year has a mean of 2 and a variance of 2. Suppose that the number of claims filed by the policyholders are independent. Approximate the probability that the total number of claims received by the insurance company during the year is between 2,450 and 2,600.

***

2

Frequency Models

In this chapter, we introduce distributions that are suitable for modeling claim frequency. Since claim frequency is measured by nonnegative integers, we need to select distributions that are discrete and have nonnegative supports. These distributions are referred to as count distributions.

2.1 The Binomial Distribution

The binomial distribution is one of the oldest count distributions that was derived by James Bernoulli before 1713 [11]. It can be used to model the number of claims occurring from a total of m independent and identical risks. The probability mass function of the binomial distribution is defined in Definition 2.1.

Definition 2.1 (Binomial distribution)**.** The probability mass function of the binomial distribution is given by:

$$p_k = \binom{m}{k} q^k (1-q)^{m-k}, \quad k = 0, 1, \ldots, m,$$

where m is a positive integer and $q \in (0, 1)$. The parameters m and q are usually referred to as the number of trials and the success probability for each trial, respectively.

If a random variable N follows the binomial distribution with parameters m and q, we write $N \sim B(m, q)$.

Figure 2.1 shows the probability mass functions of the binomial distribution with different values of the parameter q. In the plot, the parameter m is fixed to be 100. We can see that when $q = 0.5$, the probability mass function is symmetric. In other cases, the distribution is skewed. When $q < 0.5$, the binomial distribution has positive skewness (see Exercise 2.3). In this case, the tail is on the right side of the distribution. When $q > 0.5$, the binomial distribution has negative skewness. In this case, the tail is on the left side of the distribution.

DOI: 10.1201/9781003484899-2

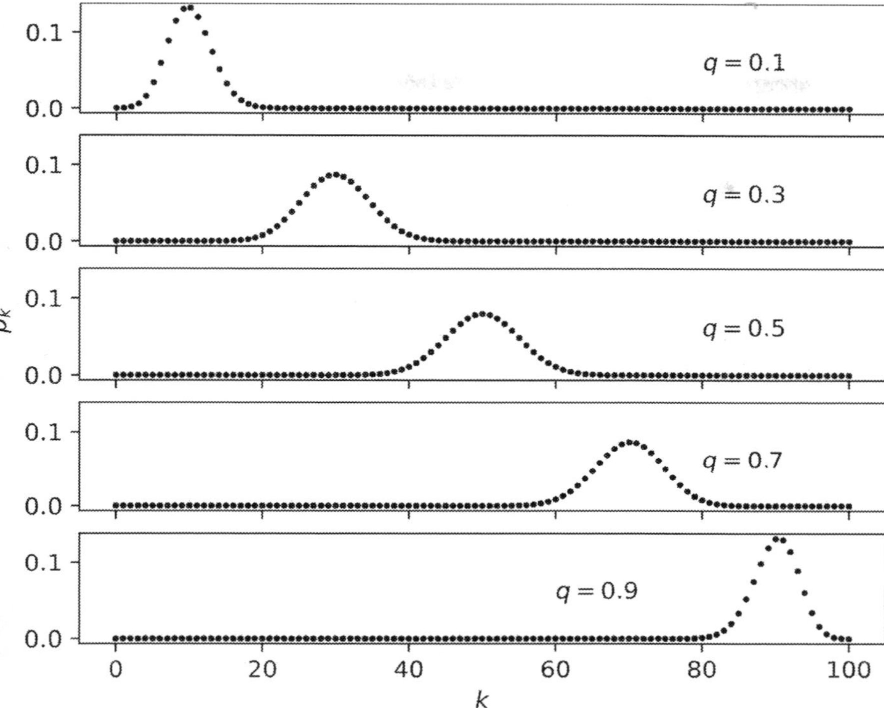

FIGURE 2.1
Probability mass functions of the binomial distribution for different values of the parameter q. The value of the parameter is 100.

The binomial distribution has some useful properties. First, it has finite support. it has positive probabilities only on the $m+1$ integers: 0, 1, ..., m. It is useful to model the number of claims when there is an upper limit. Second, its variance is smaller than its mean as is shown in Theorem 2.1. As a result, it is suitable to model count data in which the sample variance is less than the sample mean.

Theorem 2.1. *Let $N \sim B(m, q)$. Then the mean and variance of N are given by:*

$$E[N] = mq, \quad \mathrm{Var}(N) = mq(1-q).$$

Proof. The probability generating function of the binomial distribution is (see Exercise 2.1):

$$P(z) = [1 + q(z-1)]^m .$$

Hence

$$E[N] = \left.\frac{\mathrm{d}\,P(z)}{\mathrm{d}\,z}\right|_{z=1} = m\left[1+q(z-1)\right]^{m-1} q\Big|_{z=1} = mq.$$

Similarly, we have

$$\begin{aligned} E\left[N(N-1)\right] &= \left.\frac{\mathrm{d}^2\,P(z)}{\mathrm{d}\,z^2}\right|_{z=1} = m(m-1)\left[1+q(z-1)\right]^{m-2} q^2\Big|_{z=1} \\ &= m(m-1)q^2, \end{aligned}$$

which gives

$$E\left[N^2\right] = E\left[N(N-1)\right] + E[N] = m(m-1)q^2 + mq.$$

The variance is given by:

$$\mathrm{Var}(N) = E\left[N^2\right] - E[N]^2 = m(m-1)q^2 + mq - (mq)^2 = mq(1-q).$$

□

Remark 2.1. In the proof of Theorem 2.1, we used the following result:

$$E\left[N(N-1)\right] = \left.\frac{\mathrm{d}^2\,P(z)}{\mathrm{d}\,z^2}\right|_{z=1}.$$

In general, we have

$$E\left[N(N-1)(N-2)\cdots(N-k+1)\right] = \left.\frac{\mathrm{d}^k\,P(z)}{\mathrm{d}\,z^k}\right|_{z=1}.$$

The use of the probability generating function is different from that of the moment generating function.

***

Exercise 2.1. Let $N \sim B(m,q)$, where m is a positive integer and $q \in (0,1)$. Show that the probability generating function of N is given by:

$$P(z) = \left[1+q(z-1)\right]^m.$$

Exercise 2.2. Let $N \sim B(m,q)$. Show that the moment generating function of N is

$$M_N(t) = \left[1+q(e^t-1)\right]^m.$$

Exercise 2.3. Let $N \sim B(m,q)$. Show that the skewness of N is

$$\frac{1-2q}{\sqrt{mq(1-q)}}.$$

Exercise 2.4. Cities J and K are neighbors. The number of severe storms in cite J in a year follows a binomial distribution with $m = 5$ and $q = 0.6$. Given that cite J has j severe storms in a year, the number of severe storms in city K in the same year has the following distribution:

$$p_j = \frac{1}{2}, \quad p_{j+1} = \frac{1}{3}, \quad p_{j+2} = \frac{1}{6}.$$

Calculate the expected number of severe storms in city J in a year given that 5 severe storms strike city K in the same year.

Exercise 2.5. In a week, the number of days of measurable rain in an area follows a binomial distribution with $m = 7$ and $q = 0.6$. A weather insurance policy pays for \$1,000 per day, up to two days, for days of measurable rain in a week. Calculate the expected payment of the weather insurance policy.

Exercise 2.6. Let N be the number of claims from a portfolio of policies. Let $0 < p < 1$. Suppose that with probability p, N has a binomial distribution with parameters $m = 2$ and $q = 0.5$; with probability $1 - p$, N has a binomial distribution with parameters $m = 4$ and $q = 0.5$. Calculate $P(N = 2)$ by expressing it as a function of p.

***

2.2 The Poisson Distribution

The Poisson distribution is a discrete distribution named after the French mathematician Siméon Denis Poisson. It was first introduced by Poisson in 1837. The probability mass function of the Poisson distribution is given in Definition 2.2.

Definition 2.2 (Poisson distribution)**.** The probability mass function of the Poisson distribution is given by:

$$p_k = \frac{\lambda^k e^{-\lambda}}{k!}, \quad k = 0, 1, 2, \ldots,$$

where $\lambda > 0$ is a parameter.

If a random variable N follows the Poisson distribution with parameter λ, we write $N \sim \text{Pois}(\lambda)$.

Figure 2.2 shows the probability mass functions of the Poisson distribution with different values of the parameter λ. From the figure, we see that when λ is small, the distribution is skewed to the right side. When λ is larger, the

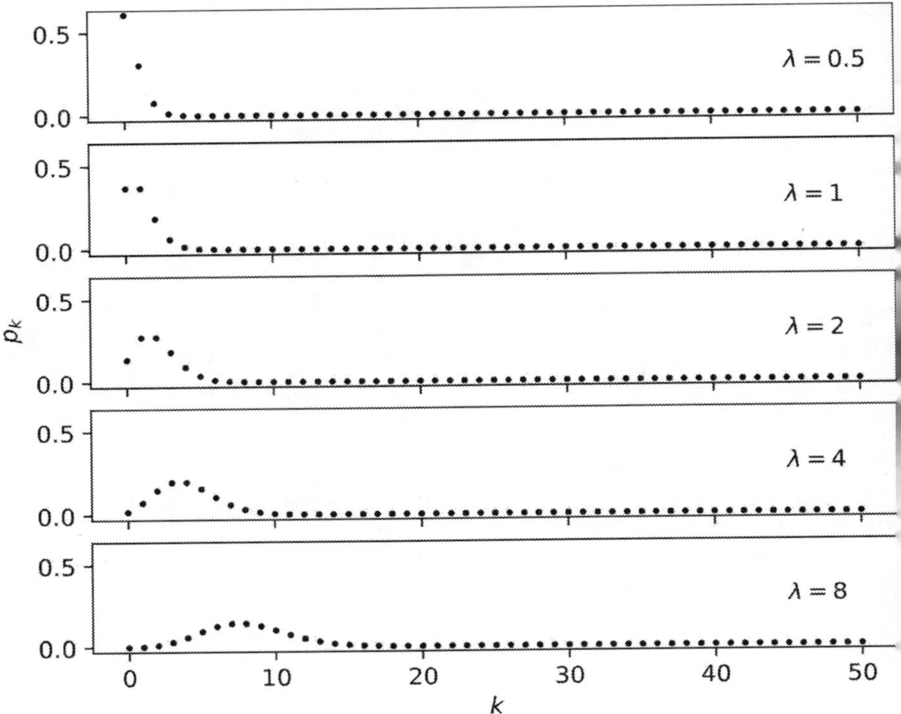

FIGURE 2.2
Probability mass functions of the Poisson distribution. The values of the parameter λ are given the subplots.

distribution becomes more symmetric. However, the skewness of the Poisson distribution is always positive (see Exercise 2.9).

The Poisson distribution has two important properties. First, its mean and its variance are equal. This is shown in Theorem 2.2. Second, the sum of independent Poisson random variables is still a Poisson random variable. The second property is shown in Theorem 2.3.

Theorem 2.2. *Let $N \sim \text{Pois}(\lambda)$. Then the mean and variance of N are given by:*

$$E[N] = \lambda, \quad \text{Var}(N) = \lambda.$$

Proof. The probability generating function of N is (see Exercise 2.7):

$$P(z) = e^{\lambda(z-1)}.$$

The mean can be calculated from the pgf as follows:

$$E[N] = P'(1) = \lambda e^{\lambda(z-1)}\Big|_{z=1} = \lambda.$$

The variance can be calculated from the pdg as follows:

$$\begin{aligned}\operatorname{Var}(N) &= E\left[N^2\right] - E[N]^2 = E[N(N-1)] + E[N] - E[N]^2 \\ &= P''(1) + E[N] - E[N]^2 = \lambda^2 e^{\lambda(z-1)}\Big|_{z=1} + \lambda - \lambda^2 \\ &= \lambda.\end{aligned}$$

In the above equation, we used the following property of the pdf: $P''(1) = E[N(N-1)]$. □

Theorem 2.3. *Let $N_i \sim \text{Pois}(\lambda_i)$ for $i = 1, 2, \ldots, n$, where $\lambda_i > 0$. Suppose that N_1, N_2, ..., N_n are independent. Then*

$$N \sim \text{Pois}(\lambda_1 + \lambda_2 + \cdots + \lambda_n),$$

where $N = N_1 + N_2 + \cdots + N_n$.

Proof. Since the n Poisson random variables are independent, we have

$$\begin{aligned}P_N(z) &= E\left[z^N\right] = E\left[z^{N_1+N_2+\cdots+N_n}\right] \\ &= \prod_{i=1}^{n} E\left[z^{N_i}\right] = \prod_{i=1}^{n} e^{\lambda_i(z-1)} \\ &= e^{(\lambda_1+\lambda_2+\cdots+\lambda_n)(z-1)}.\end{aligned}$$

Since the pgf of N is the pgf of the Poisson distribution, N follows the Poisson distribution with parameter $\lambda_1 + \lambda_2 + \cdots + \lambda_n$. □

***

Exercise 2.7. Let $N \sim \text{Pois}(\lambda)$, where $\lambda > 0$. Show that the probability generating function of N is

$$P(z) = e^{\lambda(z-1)}.$$

Exercise 2.8. Let $N \sim \text{Pois}(\lambda)$, where $\lambda > 0$. Show that the moment generating function of N is

$$M(t) = e^{\lambda(e^t-1)}.$$

Exercise 2.9. Let $N \sim \text{Pois}(\lambda)$, where $\lambda > 0$. Show that the skewness of N is

$$\frac{1}{\sqrt{\lambda}}.$$

Exercise 2.10. Show that

$$\lim_{m\to\infty, q\to 0, mq\to\lambda} \binom{m}{k} q^k (1-q)^{m-k} = \frac{\lambda^k e^{-\lambda}}{k!}, \quad k = 0, 1, \ldots.$$

(Hint: use Theorem B.3 and Theorem A.6.)

Exercise 2.11. The number of claims received by an insurance company in a year follows a Poisson distribution. It has been observed that the probability of receiving two claims is three times the probability of receiving four claims. Calculate the variance of the number of claims received by the insurance company in the year.

Exercise 2.12. The number of claims N received by an insurance company in a year follows a Poisson distribution $\text{Pois}(\Lambda)$. The parameter Λ follows a uniform distribution on $[0, 2]$. Calculate $E[N]$ and $\text{Var}(N)$.

***

2.3 The Negative Binomial Distribution

The negative binomial distribution is commonly used to model claim counts in insurance. It has two parameters and can provide more flexibility than the Poisson distribution, which has only one parameter. The probability mass function of the negative binomial distribution is given in Definition 2.3.

Definition 2.3 (Negative binomial distribution)**.** The probability mass function of the negative binomial distribution is given by:

$$p_k = \binom{k+r-1}{k} p^r (1-p)^k, \quad k = 0, 1, 2, \ldots,$$

where $r > 0$ and $p \in (0, 1)$ are two parameters.

If a random variable N follows the negative binomial distribution with parameters r and p, we write $N \sim \text{NB}(r, p)$.

The probability mass function of the negative binomial distribution is derived from the Taylor expansion of the function $(1-w)^{-r}$ at 0:

$$\begin{aligned}(1-w)^{-r} =& 1 + rw + \frac{(r+1)r}{2!} w^2 + \frac{(r+2)(r+1)r}{3!} w^3 + \cdots \\ =& \sum_{k=0}^{\infty} \binom{r+k-1}{k} w^k.\end{aligned}$$

The above series is called a negative binomial series. Letting $w = 1 - p$ in the above series, we get

$$p^{-r} = \sum_{k=0}^{\infty} \binom{r+k-1}{k} (1-p)^k,$$

which shows that

$$\sum_{k=0}^{\infty} \binom{r+k-1}{k} p^r (1-p)^k = 1.$$

In the probability mass function of the negative binomial distribution, the binomial coefficient is evaluated as follows:

$$\binom{k+r-1}{k} = \frac{(k+r-1)(k+r-2)\cdots r}{k!},$$

where r can be any positive real number and k must be a nonnegative integer. It can also be evaluated by using the gamma functions:

$$\binom{k+r-1}{k} = \frac{\Gamma(k+r)}{\Gamma(k+1)\Gamma(r)}.$$

Figure 2.3 shows the probability mass functions of the negative binomial distribution with different parameters. When the parameter r is fixed, increasing the parameter p reduces the right tail. When the parameter p is fixed, increasing the parameter r increases the right tail.

Theorem 2.4. *Let* $N \sim \mathrm{NB}(r, p)$. *Then*

$$E[N] = \frac{r(1-p)}{p}, \quad \mathrm{Var}(N) = \frac{r(1-p)}{p^2}.$$

Proof. We can calculate the mean directly as follows:

$$\begin{aligned}
E[N] &= \sum_{k=0}^{\infty} k \frac{\Gamma(k+r)}{\Gamma(k+1)\Gamma(r)} p^r (1-p)^k = \sum_{k=1}^{\infty} \frac{\Gamma(k+r)}{\Gamma(k)\Gamma(r)} p^r (1-p)^k \\
&= \sum_{k=0}^{\infty} \frac{\Gamma(k+r+1)}{\Gamma(k+1)\Gamma(r)} p^r (1-p)^{k+1} \\
&= \frac{r(1-p)}{p} \sum_{k=0}^{\infty} \frac{\Gamma(k+r+1)}{\Gamma(k+1)\Gamma(r+1)} p^{r+1} (1-p)^k \\
&= \frac{r(1-p)}{p}.
\end{aligned}$$

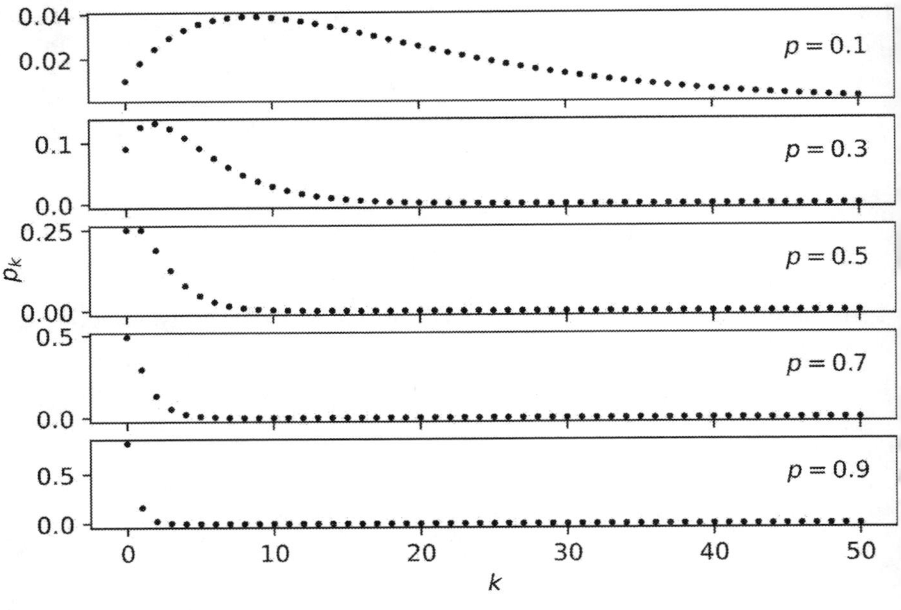

(a) The parameter r is fixed to 2.

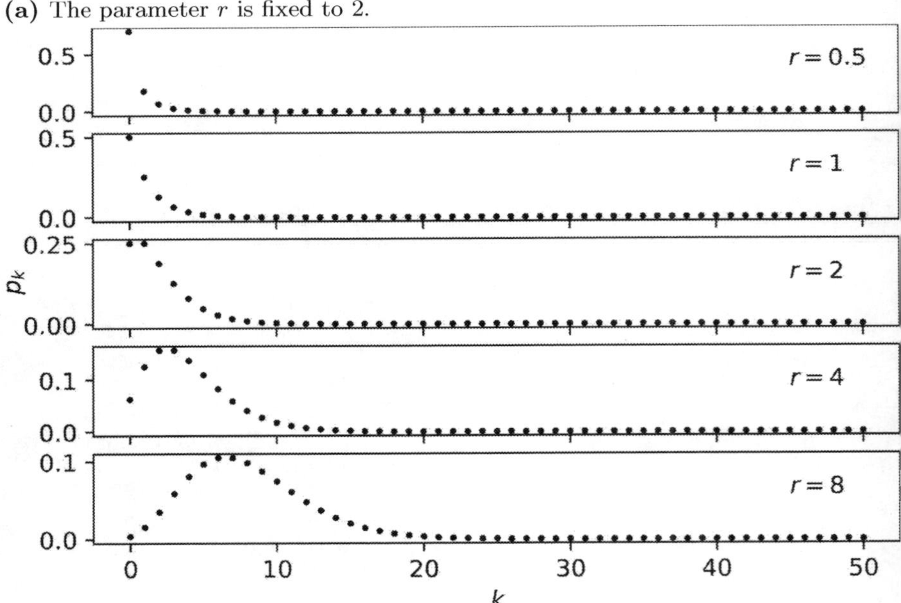

(b) The parameter p is fixed to 0.5.

FIGURE 2.3
Probability mass functions of the negative binomial distribution with different values of the parameters.

To get the variance, we need the second raw moment. However, we do not calculate $E[N^2]$ directly. Instead, we calculate $E[N(N-1)]$. This is done as follows:

$$\begin{aligned}
E[N(N-1)] &= \sum_{k=0}^{\infty} k(k-1)\frac{\Gamma(k+r)}{\Gamma(k+1)\Gamma(r)}p^r(1-p)^k \\
&= \sum_{k=2}^{\infty} \frac{\Gamma(k+r)}{\Gamma(k-1)\Gamma(r)}p^r(1-p)^k = \sum_{k=0}^{\infty} \frac{\Gamma(k+r+2)}{\Gamma(k+1)\Gamma(r)}p^r(1-p)^{k+2} \\
&= \frac{(r+1)r(1-p)^2}{p^2}\sum_{k=0}^{\infty} \frac{\Gamma(k+r+2)}{\Gamma(k+1)\Gamma(r+2)}p^{r+2}(1-p)^k \\
&= \frac{(r+1)r(1-p)^2}{p^2}.
\end{aligned}$$

Then the variance can be obtained by

$$\begin{aligned}
\mathrm{Var}(N) =& E[N^2] - E[N]^2 = E[N(N-1)] + E[N] - E[N]^2 \\
=& \frac{(r+1)r(1-p)^2}{p^2} + \frac{r(1-p)}{p} - \left(\frac{r(1-p)}{p}\right)^2 \\
=& \frac{r(1-p)}{p^2}.
\end{aligned}$$

□

From Theorem 2.4 we see that the variance of the negative binomial distribution is greater than the mean. The negative binomial distribution is suitable for data where the sample variance is greater than the sample mean.

The geometric distribution is a special case of the negative binomial distribution. When $r = 1$, the negative binomial distribution becomes the geometric distribution. The Poisson distribution is a limiting case of the negative binomial distribution (see Exercise 2.15).

***

Exercise 2.13. Let $N \sim \mathrm{NB}(r, p)$, where $r > 0$ and $p \in (0, 1)$. Show that the probability generating function of N is

$$P(z) = p^r[1 - (1-p)z]^{-r}, \quad |z| < \frac{1}{p}.$$

Exercise 2.14. Let $N \sim \mathrm{NB}(r, p)$, where $r > 0$ and $p \in (0, 1)$. Show that the moment generating function of N is

$$M(t) = p^r[1 - (1-p)e^t]^{-r}, \quad t < \ln\frac{1}{1-p}.$$

Exercise 2.15. Show that the Poisson distribution is a limiting case of the negative binomial distribution when $r \to \infty$, $p \to 1$, and $\dfrac{r(1-p)}{p} \to \lambda$.

TABLE 2.1
The $(a, b, 0)$ class distributions.

Distribution	a	b	p_0
Binomial	$-\frac{q}{1-q}$	$(m+1)\frac{q}{1-q}$	$(1-q)^m$
Poisson	0	λ	$e^{-\lambda}$
Negative binomial	$1-p$	$(r-1)(1-p)$	p^r

Exercise 2.16. The actuarial PhD program at a university needs to recruit 3 PhD students next year. According to past experience, an offer made to a student has a 60% chance of being accepted. Calculate the probability that the PhD program has to make more than 5 offers.

Exercise 2.17. Let S denote the prescription drug losses. Suppose that the number of claims follows a geometric distribution with mean 4 and the amount of each prescription is 40. Calculate $E[\max(S-100,0)]$.

***

2.4 Modification at Zero

The discrete distributions described in previous sections are members of the $(a, b, 0)$ class of distributions (see Definition 2.4). The distribution in the $(a, b, 0)$ class can be determined by three parameters a, b, and p_0. Table 2.1 shows these parameters for the discrete distributions described before.

Definition 2.4 ($(a, b, 0)$ class of distributions)**.** A discrete random variable is said to be a member of the $(a, b, 0)$ class of distributions if its probability function can be expressed as

$$p_k = \left(a + \frac{b}{k}\right) p_{k-1}, \quad k = 1, 2, \ldots,$$

where a and b are parameters. The probability p_0 can be determined by the fact that the sum of all probabilities equals to 1.

The discrete distributions may not fit the insurance data well due to the large probability at zero. In insurance data, there usually exists large number of zeros. To fit this kind of data well, we can modify the probability at zero. These modified distributions form the $(a, b, 1)$ class of distributions (see Definition 2.5).

Definition 2.5 ((a, b, 1) class of distributions)**.** A discrete random variable is said to be a member of the $(a, b, 1)$ class of distributions if its probability function $\{p_k^M\}_{k\geq 0}$ can be expressed as

$$p_k^M = \left(a + \frac{b}{k}\right) p_{k-1}^M, \quad k = 2, 3, \ldots,$$

where a, b, and p_0^M are parameters. The probability p_1 can be determined from the fact that the sum of all probabilities equals to 1.

If $p_0^M = 0$, the resulting distribution is called a zero-truncated distribution. If $p_0^M > 0$, the resulting distribution is called a zero-modified distribution.

Theorem 2.5 gives the relationship between the probability function of a $(a, b, 1)$ distribution and that of a $(a, b, 0)$ distribution. If $p_0^M > p_0$, the resulting zero-modified distribution is called a zero-inflated distribution.

Theorem 2.5. *Let $\{p_k\}_{k\geq 0}$ be the probability function of a $(a, b, 0)$ distribution. Let $\{p_k^M\}_{k\geq 0}$ be the corresponding $(a, b, 1)$ distribution, where p_0^M is given. Then for $k \geq 1$,*

$$p_k^M = \frac{1 - p_0^M}{1 - p_0} p_k.$$

Proof. Since $\{p_k\}_{k\geq 0}$ is the probability function of a $(a, b, 0)$ distribution, we have

$$p_k = \left(a + \frac{b}{k}\right) p_{k-1}, \quad k = 1, 2, \ldots.$$

By applying the above equation recursively, we get

$$p_k = p_0 \prod_{j=1}^{k} \left(a + \frac{b}{j}\right), \quad k = 1, 2, \ldots.$$

Similarly for the $(a, b, 1)$ distribution, we have

$$p_k^M = p_1^M \prod_{j=2}^{k} \left(a + \frac{b}{j}\right), \quad k = 2, 3, \ldots.$$

From the above two equations, we get

$$p_k^M = \frac{p_1^M}{p_0(a + b)} p_k, \quad k = 2, 3, \ldots. \tag{2.1}$$

Since $p_0^M + p_1^M + \cdots = 1$, we have

$$\begin{aligned}
1 =& p_0^M + p_1^M + \sum_{k=2}^{\infty} \frac{p_1^M}{p_0(a+b)} p_k = p_0^M + p_1^M + \frac{p_1^M}{p_0(a+b)} \sum_{k=2}^{\infty} p_k \\
=& p_0^M + p_1^M + \frac{p_1^M}{p_0(a+b)}(1 - p_0 - p_1) \\
=& p_0^M + p_1^M + \frac{p_1^M}{p_0(a+b)}(1 - p_0 - p_0(a+b)) = p_0^M + \frac{p_1^M}{p_0(a+b)}(1 - p_0),
\end{aligned}$$

which gives

$$p_1^M = \frac{1 - p_0^M}{1 - p_0} p_0(a+b) = \frac{1 - p_0^M}{1 - p_0} p_1. \tag{2.2}$$

Plugging Equation (2.2) into Equation (2.1), we get

$$p_k^M = \frac{1 - p_0^M}{1 - p_0} p_k, \quad k = 2, 3, \ldots.$$

From Equation (2.2), we know that the above equation also holds when $k = 1$. This finishes the proof. □

Example 2.1. Let N follow a zero-modified Poisson distribution with $p_0^M = 0.5$ and $\lambda = 2$. Calculate $P(N = 5)$.

Solution. By Theorem 2.5, we have

$$P(N = 5) = p_5^M = \frac{1 - p_0^M}{1 - p_0} p_5 = \frac{1 - 0.5}{1 - e^{-2}} \cdot \frac{2^5 e^{-2}}{5!} = \frac{2e^{-2}}{15(1 - e^{-2})}.$$

□

***

Exercise 2.18. Let $\{p_k\}_{k \geq 0}$ and $P(z)$ be the probability function and the probability generating function of a $(a, b, 0)$ distribution. Show that the probability generating function of the corresponding zero-modified distribution is given by

$$P^M(z) = \frac{p_0^M - p_0}{1 - p_0} + \frac{1 - p_0^M}{1 - p_0} P(z).$$

Exercise 2.19. Consider a random variable following a zero-truncated negative binomial distribution with parameters r and p. Calculate p_1^T.

Exercise 2.20. Let N be a random variable that follows a zero modified Poison distribution with parameters $\lambda = 3$ and $p_0^M = 0.5$. Calculate p_1^M, p_2^M, p_3^M, $E[N]$, and $\text{Var}(N)$.

Exercise 2.21. Let N be a discrete random variable from the $(a, b, 0)$ class of distributions. The following information is known about the distribution:

$$\begin{aligned} P(N = 0) &= 0.327680 \\ P(N = 1) &= 0.327680 \\ P(N = 2) &= 0.196608 \\ E[N] &= 1.25. \end{aligned}$$

Calculate $P(N = 3)$ and determine the name of the distribution of N.

Exercise 2.22. The pf of a discrete random variable satisfies the following recursive formula:

$$p_k = \frac{2}{k} p_{k-1}, \quad k = 1, 2, \ldots.$$

Calculate p_4.

---***

3

Severity Models

In this chapter, we introduce distributions that are suitable for modeling claim severity. Since claim severity is usually positive, we need to select distributions that are continuous and have positive support. These distributions are also called size distributions.

3.1 The Linear Exponential Family

The linear exponential family is a family of distributions that include many common discrete and continuous distributions. Definition 3.1 gives the definition of the linear exponential family.

Definition 3.1 (Linear exponential family)**.** A distribution is said to be a member of the linear exponential family if its pdf can be expressed as

$$f(x;\theta) = \frac{p(x)e^{r(\theta)x}}{q(\theta)}, \tag{3.1}$$

where θ is a parameter, $p(x)$ is a function depending only on x, $q(\theta)$ is a function depending only on θ, and $r(\theta)$ is also a function depending only on θ. The support of the distribution should not depend on θ. The function $q(\theta)$ is a normalizing factor and $r(\theta)$ is called the canonical or natural parameter of the distribution.

The normal distribution is a member of the linear exponential family. Example 3.1 shows the details. There are distributions that do not belong to the linear exponential family. See Example 3.2.

Example 3.1. The density of the normal distribution $N(\mu, \sigma^2)$ is given by

$$f(x) = \frac{1}{\sqrt{2\pi}\sigma} \exp\left(-\frac{(x-\mu)^2}{2\sigma^2}\right).$$

Show that the normal distribution is a member of the linear exponential family.

DOI: 10.1201/9781003484899-3

Solution. This can be shown by rewriting the density function as follows:

$$\begin{aligned} f(x) &= \frac{1}{\sqrt{2\pi}\sigma} \exp\left(-\frac{x^2 - 2\mu x + \mu^2}{2\sigma^2}\right) \\ &= \exp\left(-\frac{x^2}{2\sigma^2}\right) \frac{1}{\sqrt{2\pi}\sigma \exp\left(\frac{\mu^2}{2\sigma^2}\right)} \exp\left(\frac{\mu}{\sigma^2}x\right). \end{aligned}$$

Let $p(x) = \exp\left(-\frac{x^2}{2\sigma^2}\right)$, $q(\mu) = \sqrt{2\pi}\sigma \exp\left(\frac{\mu^2}{2\sigma^2}\right)$, and $r(\mu) = \frac{\mu}{\sigma^2}$. Then $f(x)$ becomes the form given in Equation (3.1). □

Example 3.2. Consider the following shifted exponential distribution:

$$F(x) = 1 - e^{-x+\theta}, \quad x > \theta,$$

where θ is a real number. Show that it is not a member of the linear exponential family.

Solution. Since the support of this shifted exponential distribution depends on the parameter θ, it cannot be a member of the linear exponential family. □

The mean and the variance of distributions in the linear exponential family can be calculated conveniently from the functions $r(\theta)$ and $q(\theta)$. Theorem 3.1 gives the formulas to calculate the mean and the variance in terms of the function $r(\theta)$ and $q(\theta)$.

Theorem 3.1. *Let X be a random variable following a distribution from the linear exponential family. Then the mean and the variance of X are given by:*

$$E[X] = \mu(\theta) = \frac{q'(\theta)}{r'(\theta)q(\theta)},$$

$$\operatorname{Var}(X) = \frac{\mu'(\theta)}{r'(\theta)}.$$

Proof. Since the distribution of X is a member of the linear exponential family, the density of X has the form given in Equation (3.1). The log density is given by:

$$\ln f(x;\theta) = \ln p(x) + r(\theta)x - \ln q(\theta).$$

Taking derivatives of both side with respect to θ gives:

$$\frac{1}{f(x;\theta)} \frac{\partial f(x;\theta)}{\mathrm{d}\,\theta} = r'(\theta)x - \frac{q'(\theta)}{q(\theta)},$$

which can be rearranged as:

$$\frac{\partial f(x;\theta)}{\mathrm{d}\,\theta} = \left(r'(\theta)x - \frac{q'(\theta)}{q(\theta)}\right) f(x;\theta). \tag{3.2}$$

Integrating both sides of the above equation over the range of x, we get

$$\int_S \frac{\partial f(x;\theta)}{\mathrm{d}\,\theta}\,\mathrm{d}\,x = r'(\theta)\int_S x f(x;\theta)\,\mathrm{d}\,x - \frac{q'(\theta)}{q(\theta)}\int_S f(x;\theta)\,\mathrm{d}\,x,$$

where S is the support of X. Note that $\int_S f(x;\theta)\,\mathrm{d}\,x = 1$ and $\int_S x f(x;\theta)\,\mathrm{d}\,x = E[X]$. By changing the order of integration and differentiation[1], we have

$$\int_S \frac{\partial f(x;\theta)}{\mathrm{d}\,\theta}\,\mathrm{d}\,x = \frac{\partial}{\mathrm{d}\,\theta}\int_S f(x;\theta)\,\mathrm{d}\,x = \frac{\partial}{\mathrm{d}\,\theta}(1) = 0.$$

Combining the above results, we get

$$E[X] = \frac{q'(\theta)}{r'(\theta)q(\theta)}.$$

Let $\mu(\theta) = \dfrac{q'(\theta)}{r'(\theta)q(\theta)}$. Equation (3.2) can be written as

$$\frac{\partial f(x;\theta)}{\mathrm{d}\,\theta} = r'(\theta)\,(x - \mu(\theta))\,f(x;\theta).$$

Taking derivatives again with respect to θ, we get

$$\begin{aligned}\frac{\partial^2 f(x;\theta)}{\mathrm{d}\,\theta^2} &= f(x;\theta)\frac{\partial}{\mathrm{d}\,\theta}\left[r'(\theta)\,(x-\mu(\theta))\right] + r'(\theta)\,(x-\mu(\theta))\,\frac{\partial f(x;\theta)}{\partial\theta}\\ &= r''(\theta)[x-\mu(\theta)]f(x;\theta) - r'(\theta)\mu'(\theta)f(x;\theta) +\\ &\quad r'(\theta)^2\,(x-\mu(\theta))^2\,f(x;\theta).\end{aligned}$$

Integrating both sides of the above equation over the support of X, we get

$$\begin{aligned}\int_S \frac{\partial^2 f(x;\theta)}{\mathrm{d}\,\theta^2}\,\mathrm{d}\,x &= r''(\theta)\int_S [x-\mu(\theta)]f(x;\theta)\,\mathrm{d}\,x - r'(\theta)\mu'(\theta)\int_S f(x;\theta)\,\mathrm{d}\,x +\\ &\quad r'(\theta)^2\int_S (x-\mu(\theta))^2\,f(x;\theta)\,\mathrm{d}\,x\end{aligned}$$

Note that $\int_S (x-\mu(\theta))^2\,f(x;\theta)\,\mathrm{d}\,x = \mathrm{Var}(X)$ and

$$\int_S \frac{\partial^2 f(x;\theta)}{\mathrm{d}\,\theta^2}\,\mathrm{d}\,x = \frac{\partial}{\mathrm{d}\,\theta}\int_S \frac{\partial f(x;\theta)}{\mathrm{d}\,\theta}\,\mathrm{d}\,x = \frac{\partial}{\mathrm{d}\,\theta}(0) = 0.$$

[1]This can be done when certain conditions are met. See Theorem 5.4.1 in [8, p224].

Combining the above results, we get

$$\mathrm{Var}(X) = \frac{\mu'(\theta)}{r'(\theta)}.$$

□

***

Exercise 3.1. The density function of the gamma distribution is given by

$$f(x) = \frac{(x/\theta)^\alpha e^{-x/\theta}}{x\Gamma(\alpha)},$$

where $\alpha > 0$ and $\theta > 0$ are parameters. Show that the gamma distribution is a member of the linear exponential family.

Exercise 3.2. Calculate the mean and the variance of the normal distribution by using Theorem 3.1.

Exercise 3.3. The Bernoulli distribution is a discrete distribution with the following probability function:

$$f(1) = P(X = 1) = \pi, \quad f(0) = P(X = 0) = 1 - \pi,$$

where X is a Bernoulli random variable and $\pi \in (0, 1)$ is a parameter. Show that the Bernoulli distribution is a member of the linear exponential family.

Exercise 3.4. A distribution of the linear exponential family has the following pdf:

$$f(x;\theta) = \frac{xe^{-\theta x}}{q(\theta)}, \quad x > 0.$$

Determine $q(\theta)$.

***

3.2 The Generalized Gamma Family

The generalized gamma family contains several common distributions that are suitable for modeling severity. The generalized gamma distribution is obtained by power transformations of the gamma distribution (see Exercise 3.5). Definitions 3.2 and 3.3 give the density functions of the transformed gamma distribution and the inverse transformed gamma distribution, respectively.

Definition 3.2 (Transformed gamma distribution)**.** The density function of the transformed gamma distribution is given by:

$$f(x) = \frac{\tau (x/\theta)^{\alpha\tau} e^{-(x/\theta)^\tau}}{x\Gamma(\alpha)}, \quad x > 0,$$

where $\alpha > 0$, $\theta > 0$, and $\tau > 0$ are parameters.

Definition 3.3 (Inverse transformed gamma distribution)**.** The density function of the inverse transformed gamma distribution is given by:

$$f(x) = \frac{\tau (\theta/x)^{\alpha\tau} e^{-(\theta/x)^\tau}}{x\Gamma(\alpha)}, \quad x > 0,$$

where $\alpha > 0$, $\theta > 0$, and $\tau > 0$ are parameters.

From Definitions 3.2 and 3.3, we see that the density functions of the transformed gamma distribution and the inverse transformed gamma distribution are similar. In fact, the two density functions can be unified as follows:

$$f(x) = \frac{|\tau| (x/\theta)^{\alpha\tau} e^{-(x/\theta)^\tau}}{x\Gamma(\alpha)}, \quad x > 0,$$

where $\alpha > 0$, $\theta > 0$, and $\tau \neq 0$.

Figures 3.1 and 3.2 show the densities of the transformed gamma distribution with different shape parameters. From the plots, we see that the shape parameters α and τ have different effects on the densities. Some densities of the inverse transformed gamma distributions are shown in Figures 3.3 and 3.4. The densities of the inverse transformed gamma distributions behave differently than those of the just inverse gamma distributions.

Figures 3.5 and 3.6 show the special cases of the transformed gamma distribution and the inverse transformed gamma distribution, respectively. The lognormal distribution is similar to the transformed gamma distribution. However, it is not a limiting case of the transformed gamma distribution (see Remark 3.1).

Remark 3.1 (Lognormal as a limiting case of the transformed gamma). It has been reported that the lognormal distribution is a limiting case of the transformed gamma distribution (see, for example, [12, p149] and [13, p77]). However, this limiting case is not the true limiting case but rather an approximation. [14] provided a derivation of the lognormal distribution as a limiting case of the transformed gamma distribution when $\alpha \to \infty$. In the derivation, the parameter α is allowed to remain constant in some places while it approaches to infinity in other places.

***

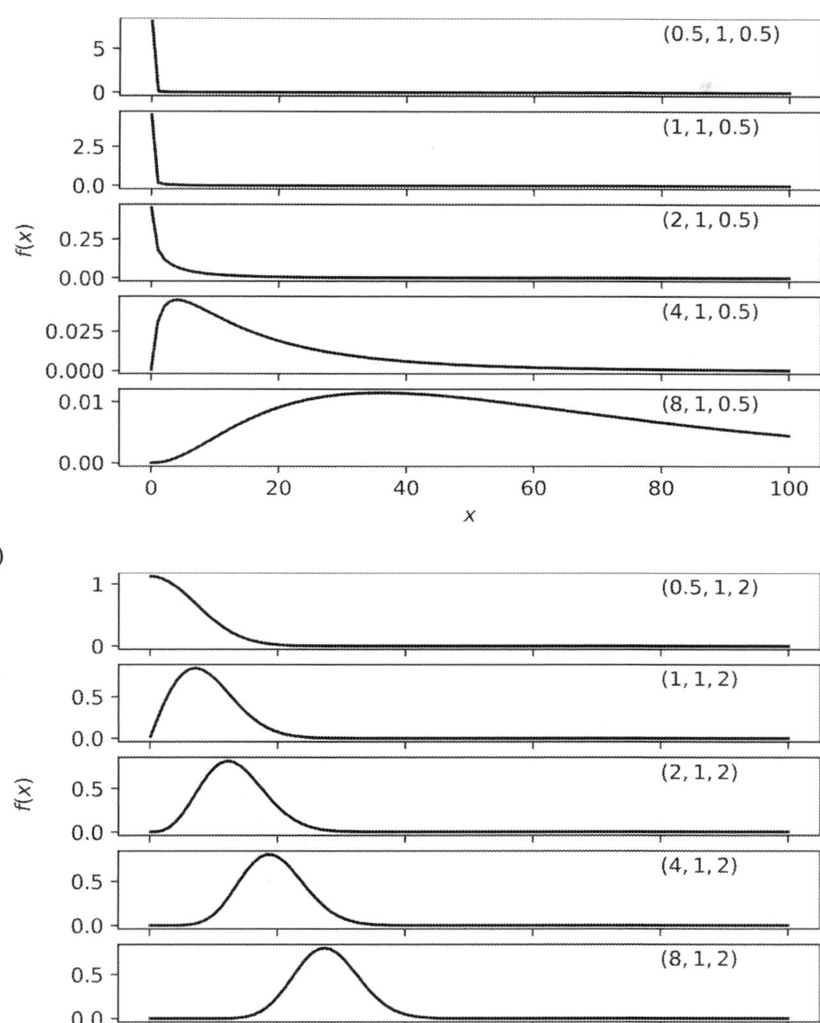

FIGURE 3.1
Densities of the transformed gamma distribution with different values of the parameter α when the parameters θ and τ are fixed. The numbers in the subplots correspond to the values of the parameters (α, θ, τ).

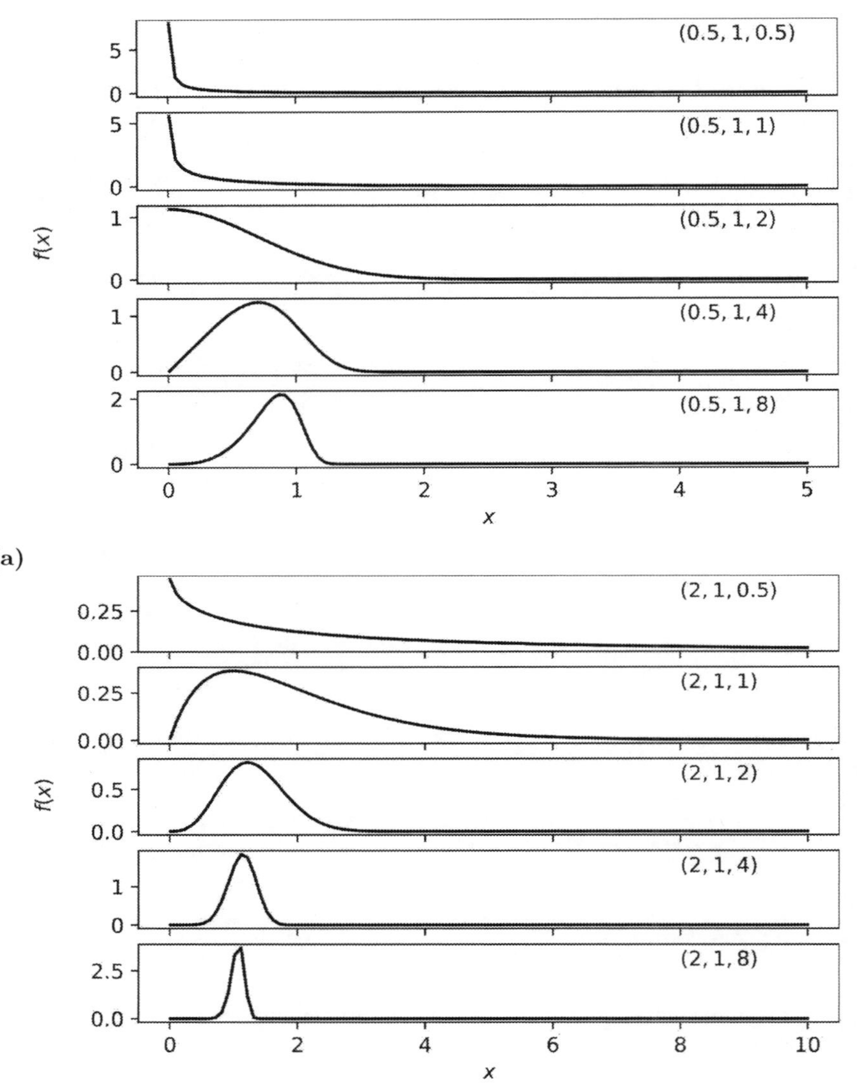

(a)

(b)

FIGURE 3.2
Densities of the transformed gamma distribution with different values of the parameter τ when the parameters θ and α are fixed. The numbers in the subplots correspond to the values of the parameters (α, θ, τ).

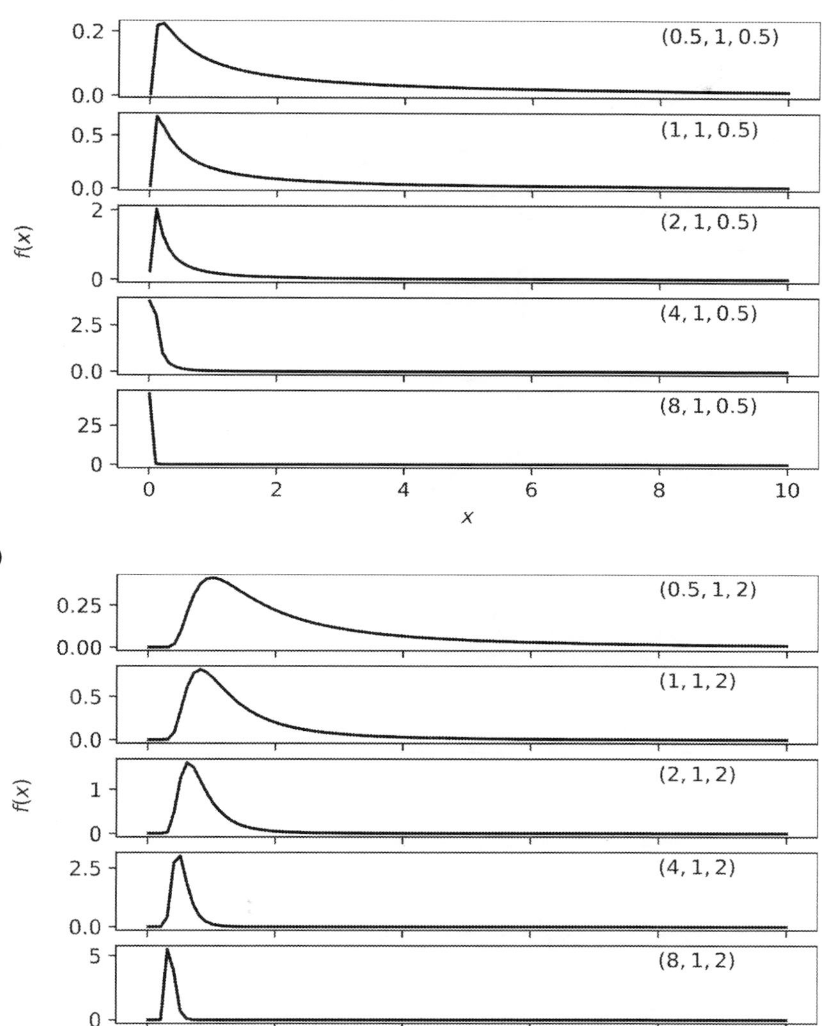

FIGURE 3.3
Densities of the inverse transformed gamma distribution with different values of the parameter α when the parameters θ and τ are fixed. The numbers in the subplots correspond to the values of the parameters (α, θ, τ).

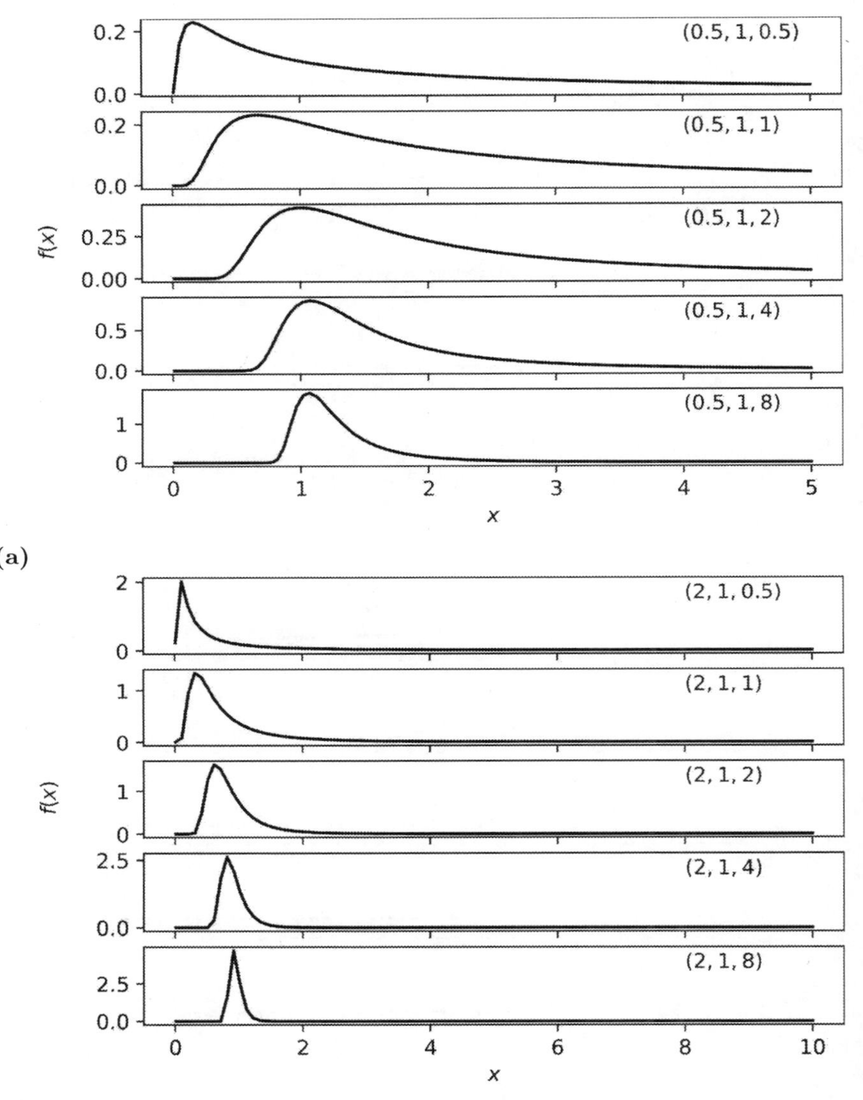

FIGURE 3.4
Densities of the inverse transformed gamma distribution with different values of the parameter τ when the parameters θ and α are fixed. The numbers in the subplots correspond to the values of the parameters (α, θ, τ).

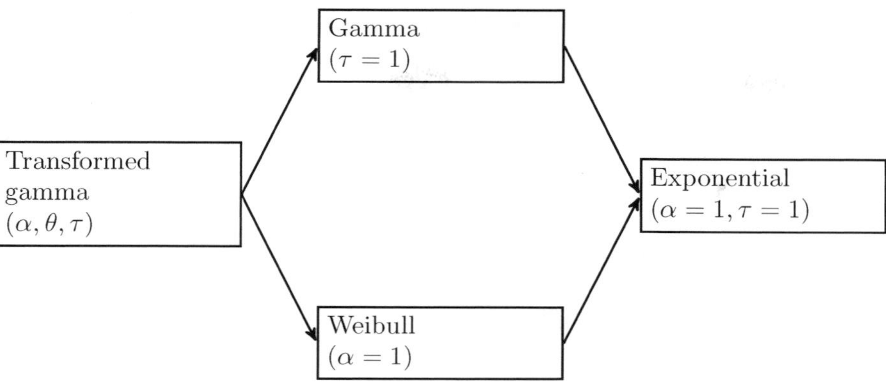

FIGURE 3.5
The transformed gamma distribution and its special cases.

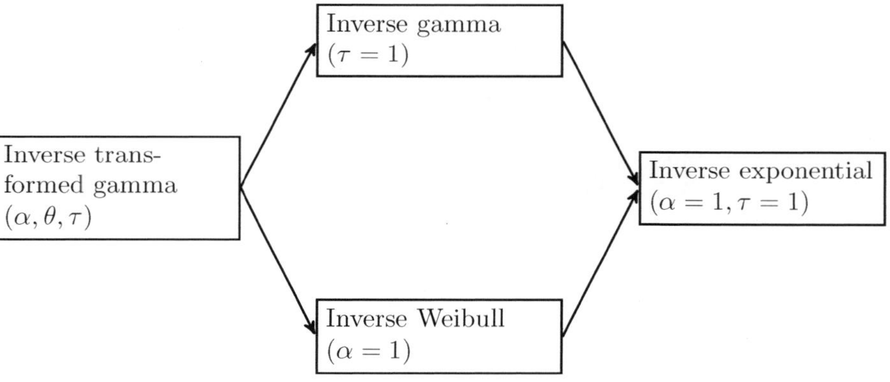

FIGURE 3.6
The inverse transformed gamma distribution and its special cases.

Exercise 3.5. Let X be a gamma random variable with the following density function:

$$f(x) = \frac{(x/\theta)^\alpha e^{-x/\theta}}{x\Gamma(\alpha)}, \quad x > 0,$$

where $\alpha > 0$ and $\theta > 0$ are parameters. Let $\tau \neq 0$. Determine the pdf of the random variable $X^{\frac{1}{\tau}}$.

Exercise 3.6. Let X be a transformed gamma random variable with the following pdf:

$$f(x) = \frac{\tau(x/\theta)^{\alpha\tau} e^{-(x/\theta)^\tau}}{x\Gamma(\alpha)}, \quad x > 0,$$

where $\alpha > 0$, $\theta > 0$, and $\tau > 0$. Calculate $E[X]$ and $\mathrm{Var}(X)$.

Exercise 3.7. Let X follow the exponential distribution with parameter θ. Let $s > 0$ and $t > 0$. Show that

$$P(X > t + s|X > s) = P(X > t).$$

Exercise 3.8. Let n be a positive integer. Show that

$$\int_1^\infty x^n e^{-x} \,\mathrm{d}\,x = e^{-1} \sum_{j=0}^{n} \frac{n!}{j!}.$$

Exercise 3.9. Compute the following integral

$$\int_1^\infty e^{-t} t^4 \,\mathrm{d}\,t.$$

Exercise 3.10. Compute the following integral

$$\int_0^\infty e^{-t} t^{5/2} \,\mathrm{d}\,t.$$

Exercise 3.11. The number of windshield claims filed per year per driver follows the Poisson distribution with parameter Λ, which is a random variable that follows the gamma distribution with mean 3 and variance 3. Calculate the probability that a randomly selected driver will file no more than 1 windshield claim next year.

---***

3.3 The GB2 Family

The GB2 (generalized beta of the second kind) distribution provides a flexible family of distributions that can be used to fit skewed data. The GB2 distribution was proposed independently by different authors. [19] proposed the GB2 distribution based on power transformations of beta and gamma distributions. [16] proposed the GB2 distribution to fit U.S. family income. [15] also proposed the GB2 distribution in a different form of parameterization as an income distribution. The GB2 distribution has been adopted in the actuarial field. For example, [5] used the GB2 distribution to model insurance loss data. [18] applied the GB2 distribution to fit heavy-tailed longitudinal data in insurance. [6] used the GB2 distribution to model the fair market values of the guarantees embedded in variable annuities.

Definition 3.4 (GB2 distribution). The density of the GB2 distribution is given by [13, 12]:

$$\begin{aligned} f(x) &= \frac{\Gamma(\alpha+\tau)}{\Gamma(\alpha)\Gamma(\tau)} \frac{\gamma(x/\theta)^{\gamma\tau}}{x[1+(x/\theta)^{\gamma}]^{\alpha+\tau}} \\ &= \frac{1}{B(\alpha,\tau)} \frac{\gamma(x/\theta)^{\gamma\tau}}{x[1+(x/\theta)^{\gamma}]^{\alpha+\tau}}, \quad x > 0, \end{aligned} \tag{3.3}$$

where $\alpha > 0$, $\gamma > 0$, $\tau > 0$ are shape parameters, $\theta > 0$ is the scale parameter, $\Gamma(\cdot)$ is the gamma function, and $B(\cdot, \cdot)$ is the beta function. If X is a GB2 random variable with the above density function, we denote

$$X \sim \mathrm{GB2}(\alpha, \theta, \gamma, \tau).$$

The density function of the GB2 distribution is given in Definition 3.4. In the definition, the parameter γ is restricted to be positive. In fact, the parameter γ can be negative. In such cases, the density function should be the absolute value of the density function given in Equation (3.3) (see Exercise 3.14).

Figures 3.7, 3.8, and 3.9 show the plots of the GB2 density function with different values of the shape parameters. In these plots, the scale parameter is set to 1. From Figures 3.7a and 3.7b, we see that larger values of α make the tail thinner.

Theorem 3.2 (GB2 moments). *Let $X \sim \mathrm{GB2}(\alpha, \theta, \gamma, \tau)$. Let k be a real number such that $-\gamma\tau < k < \gamma\alpha$. Then the kth moment of X exists and is given by:*

$$E\left[X^k\right] = \frac{\theta^k B\left(\alpha - \frac{k}{\gamma}, \tau + \frac{k}{\gamma}\right)}{B(\alpha, \tau)}. \tag{3.4}$$

Proof. Let $k \in (-\gamma\tau, \gamma\alpha)$. By definition, the kth moment is calculated by:

$$E\left[X^k\right] = \int_0^\infty x^k f(x)\,\mathrm{d}x = \int_0^\infty \frac{x^k}{B(\alpha,\tau)} \frac{\gamma(x/\theta)^{\gamma\tau}}{x[1+(x/\theta)^{\gamma}]^{\alpha+\tau}}\,\mathrm{d}x. \tag{3.5}$$

Let $u = \dfrac{1}{1+(x/\theta)^{\gamma}}$. When x changes from 0 to ∞, u will change from 1 to 0.

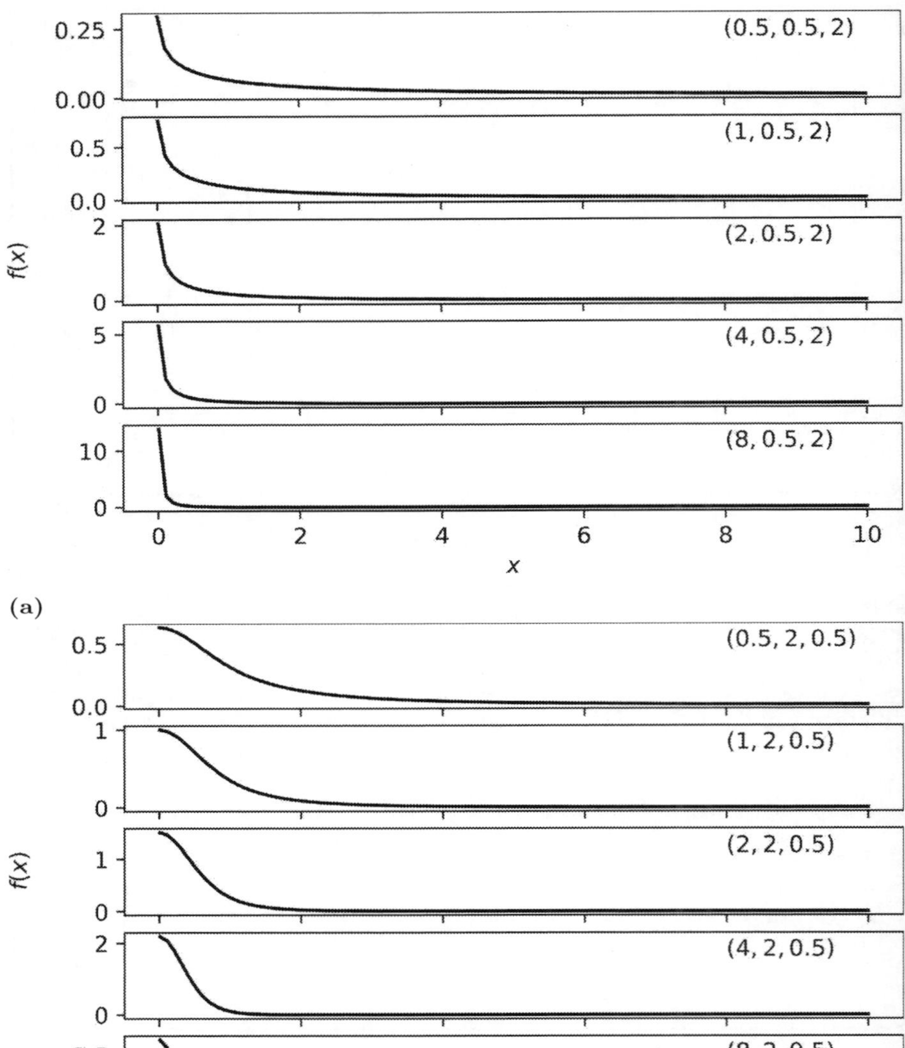

FIGURE 3.7
Densities of the GB2 distribution with different values of the parameter α when the parameters γ and τ are fixed. The numbers in the subplots correspond to the values of the parameters (α, γ, τ).

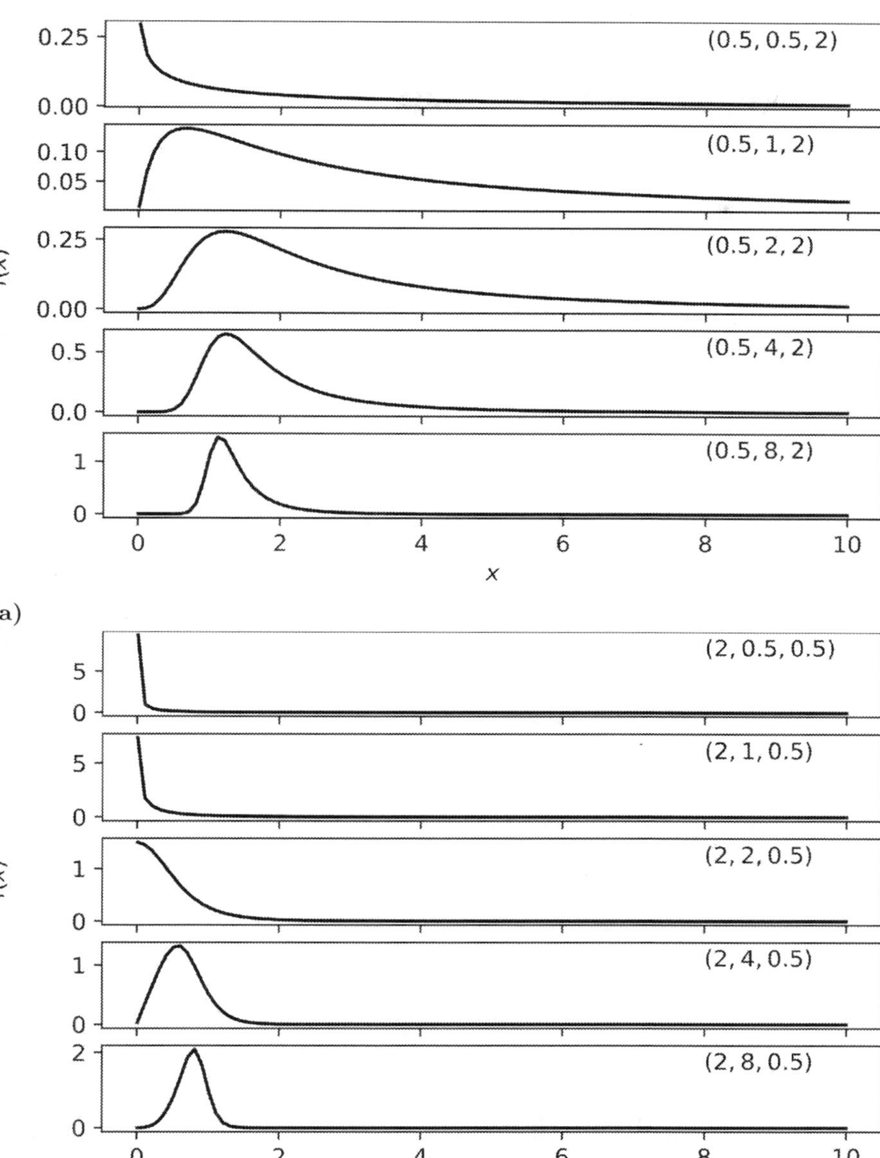

FIGURE 3.8
Densities of the GB2 distribution with different values of the parameter γ when the parameters α and τ are fixed. The numbers in the subplots correspond to the values of the parameters (α, γ, τ).

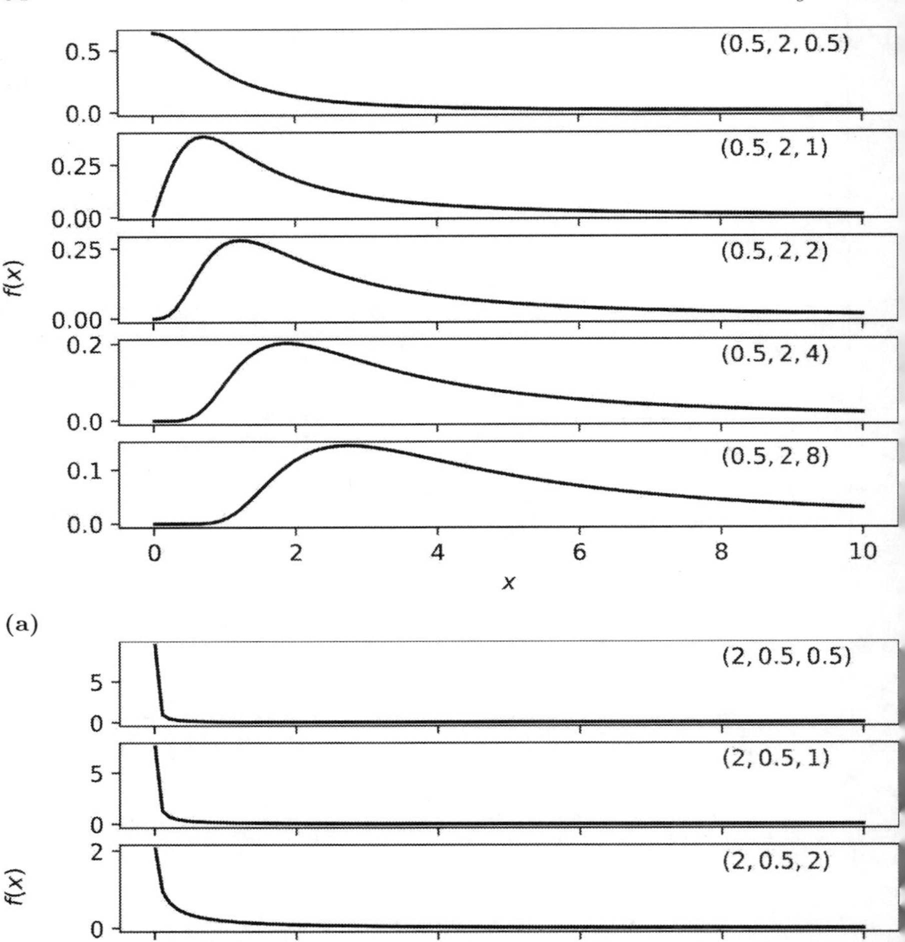

(a)

(b)

FIGURE 3.9
Densities of the GB2 distribution with different values of the parameter τ when the parameters α and γ are fixed. The numbers in the subplots correspond to the values of the parameters (α, γ, τ).

In addition, $(x/\theta)^\gamma = \dfrac{1-u}{u}$. By using the chain rule, we get

$$\begin{aligned} \mathrm{d}\,u &= -\left(1+(x/\theta)^\gamma\right)^{-2}\gamma(x/\theta)^\gamma\frac{1}{x}\,\mathrm{d}\,x = -\left(1+\frac{1-u}{u}\right)^{-2}\frac{1-u}{u}\frac{\gamma}{x}\,\mathrm{d}\,x \\ &= -u(1-u)\gamma\frac{\mathrm{d}\,x}{x} \end{aligned}$$

or

$$\frac{\gamma\,\mathrm{d}\,x}{x} = \frac{-1}{u(1-u)}\,\mathrm{d}\,u.$$

By the above change of variables and the definition of the beta function (see Definition B.2), Equation (3.5) becomes

$$\begin{aligned} E\left[x^k\right] &= \frac{\theta^k}{B(\alpha,\tau)}\int_0^\infty \frac{\gamma(x/\theta)^{\gamma\tau+k}}{x[1+(x/\theta)^\gamma]^{\alpha+\tau}}\,\mathrm{d}\,x \\ &= \frac{\theta^k}{B(\alpha,\tau)}\int_0^\infty \left(\frac{1}{1+(x/\theta)^\gamma}\right)^{\alpha+\tau}(x/\theta)^{\gamma(\tau+k/\gamma)}\frac{\gamma\,\mathrm{d}\,x}{x} \\ &= \frac{\theta^k}{B(\alpha,\tau)}\int_1^0 u^{\alpha+\tau}\left(\frac{1-u}{u}\right)^{\tau+k/\gamma}\frac{-1}{u(1-u)}\,\mathrm{d}\,u \\ &= \frac{\theta^k}{B(\alpha,\tau)}\int_0^1 u^{\alpha-k/\gamma-1}\,(1-u)^{\tau+k/\gamma-1}\,\mathrm{d}\,u \\ &= \frac{\theta^k B\left(\alpha-\frac{k}{\gamma},\tau+\frac{k}{\gamma}\right)}{B(\alpha,\tau)}. \end{aligned}$$

This finishes the proof. □

Theorem 3.2 gives the moments of a GB2 random variable. From the theorem, we see that moments of the GB2 distribution exist when the shape parameters satisfy certain conditions. For example, the mean of a GB2 random variable exists when $\gamma\alpha > 1$ and is given by:

$$E[X] = \frac{\theta B\left(\alpha-\frac{1}{\gamma},\tau+\frac{1}{\gamma}\right)}{B(\alpha,\tau)}.$$

The GB2 distribution contains many loss distributions as special cases or limiting cases. Figure 3.10 shows a diagram of special cases of the GB2 distribution. By restricting each of the shape parameters to be 1, we get three special distributions, which contain other distributions as special cases. All these special cases are obtained by restricting the shape parameters.

Figure 3.11 shows the limiting cases of the GB2 distribution. The conditions of how the parameters change are also given in the diagram. Examples 3.3 and 3.4 give the detailed derivations of the limiting cases.

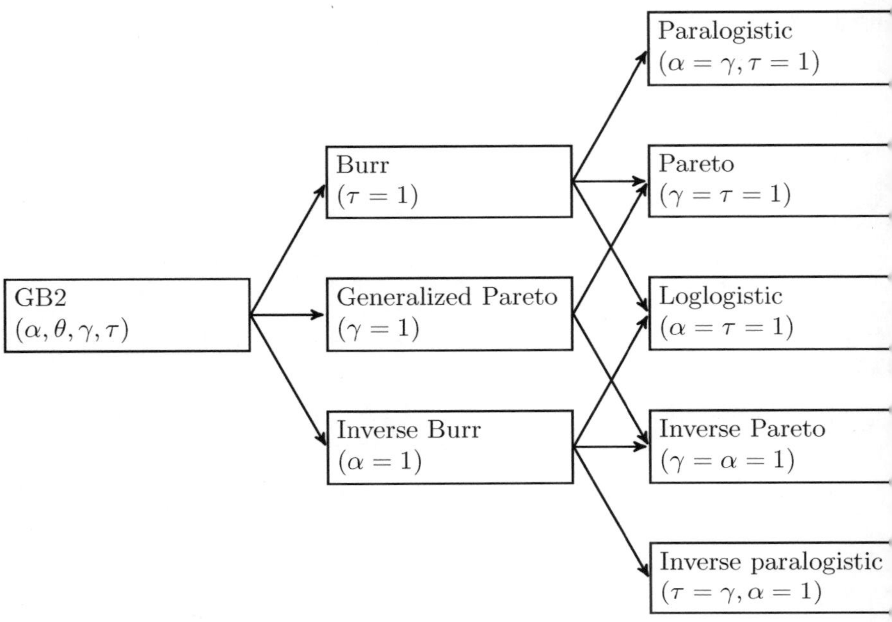

FIGURE 3.10
Special cases of the GB2 distribution.

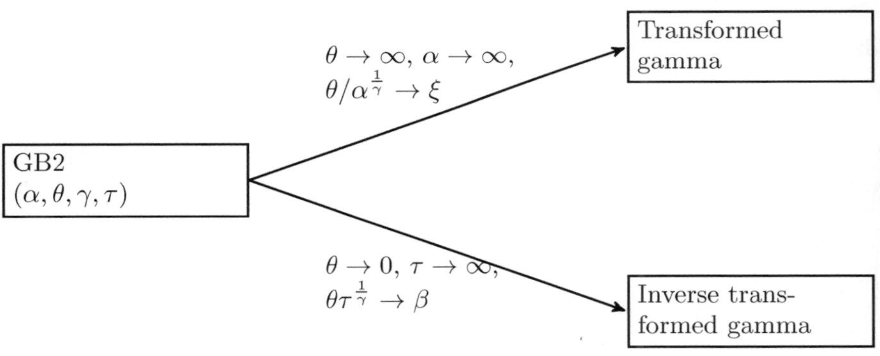

FIGURE 3.11
Limiting cases of the GB2 distribution.

Example 3.3. Let $\xi > 0$. Determine the pdf of the limiting case of the GB2 distribution when $\theta \to \infty$, $\alpha \to \infty$, and $\theta/\alpha^{\frac{1}{\gamma}} \to \xi$.

Solution. To determine the limiting pdf, we need Stirling's approximation (see Theorem B.3) and a result from calculus (i.e., Theorem A.6). To use these results, we first rewrite the GB2 density function as follows:

$$\begin{aligned}
f(x) &= \frac{\Gamma(\alpha+\tau)}{\Gamma(\alpha)\Gamma(\tau)} \frac{\gamma(x/\theta)^{\gamma\tau}}{x[1+(x/\theta)^\gamma]^{\alpha+\tau}} \\
&= \frac{\Gamma(\alpha+\tau)}{e^{-\alpha-\tau}(\alpha+\tau)^{\alpha+\tau-\frac{1}{2}}\sqrt{2\pi}} \cdot \frac{e^{-\alpha}\alpha^{\alpha-\frac{1}{2}}\sqrt{2\pi}}{\Gamma(\alpha)} \cdot \\
&\quad \frac{e^{-\alpha-\tau}(\alpha+\tau)^{\alpha+\tau-\frac{1}{2}}\sqrt{2\pi}}{e^{-\alpha}\alpha^{\alpha-\frac{1}{2}}\sqrt{2\pi}} \cdot \frac{\gamma x^{\gamma\tau-1}}{\Gamma(\tau)} \cdot \frac{1}{\theta^{\gamma\tau}\left[1+x^\gamma\theta^{-\gamma}\right]^{\alpha+\tau}} \\
&= \frac{\Gamma(\alpha+\tau)}{e^{-\alpha-\tau}(\alpha+\tau)^{\alpha+\tau-\frac{1}{2}}\sqrt{2\pi}} \cdot \frac{e^{-\alpha}\alpha^{\alpha-\frac{1}{2}}\sqrt{2\pi}}{\Gamma(\alpha)} \cdot \\
&\quad \frac{e^{-\tau}\gamma x^{\gamma\tau-1}}{\Gamma(\tau)} \cdot \frac{(\alpha+\tau)^{\alpha-\frac{1}{2}}}{\alpha^{\alpha-\frac{1}{2}}} \cdot \frac{(\alpha+\tau)^\tau}{\theta^{\gamma\tau}\left[1+x^\gamma\theta^{-\gamma}\right]^{\alpha+\tau}} \\
&= \frac{\Gamma(\alpha+\tau)}{e^{-\alpha-\tau}(\alpha+\tau)^{\alpha+\tau-\frac{1}{2}}\sqrt{2\pi}} \cdot \frac{e^{-\alpha}\alpha^{\alpha-\frac{1}{2}}\sqrt{2\pi}}{\Gamma(\alpha)} \cdot \\
&\quad \frac{e^{-\tau}\gamma x^{\gamma\tau-1}}{\Gamma(\tau)} \cdot \left(1+\frac{\tau}{\alpha}\right)^{\alpha-\frac{1}{2}} \cdot \frac{(\alpha/\theta^\gamma+\tau/\theta^\gamma)^\tau}{\left(\left[1+x^\gamma\theta^{-\gamma}\right]^{\theta^\gamma}\right)^{\alpha/\theta^\gamma+\tau/\theta^\gamma}}
\end{aligned}$$

By the assumption that $\theta \to \infty$ and $\theta/\alpha^{\frac{1}{\gamma}} \to \xi$, we have

$$\lim_{\theta\to\infty,\theta/\alpha^{\frac{1}{\gamma}}\to\xi} (\alpha/\theta^\gamma+\tau/\theta^\gamma)^\tau = \xi^{-\gamma\tau}.$$

By Theorem B.3, we have

$$\lim_{\alpha\to\infty} \frac{\Gamma(\alpha+\tau)}{e^{-\alpha-\tau}(\alpha+\tau)^{\alpha+\tau-\frac{1}{2}}\sqrt{2\pi}} = 1$$

and

$$\lim_{\alpha\to\infty} \frac{e^{-\alpha}\alpha^{\alpha-\frac{1}{2}}\sqrt{2\pi}}{\Gamma(\alpha)} = 1.$$

By Theorem A.6, we have

$$\lim_{\alpha\to\infty} \left(1+\frac{\tau}{\alpha}\right)^{\alpha-\frac{1}{2}} = e^\tau$$

and

$$\lim_{\alpha,\theta\to\infty,\theta/\alpha^{\frac{1}{\gamma}}\to\xi} \left(\left[1+x^\gamma\theta^{-\gamma}\right]^{\theta^\gamma}\right)^{\alpha/\theta^\gamma+\tau/\theta^\gamma} = e^{x^\gamma\xi^{-\gamma}}.$$

Combining the above equations, we get

$$\lim_{\alpha,\theta\to\infty,\theta/\alpha^{\frac{1}{\gamma}}\to\xi} f(x) = \frac{e^{-\tau}\gamma x^{\gamma\tau-1}}{\Gamma(\tau)} \cdot e^\tau \cdot \frac{\xi^{-\gamma\tau}}{e^{x^\gamma\xi^{-\gamma}}} = \frac{\gamma(x/\xi)^{\gamma\tau}e^{-(x/\xi)^\gamma}}{x\Gamma(\tau)},$$

which is the density function of the transformed gamma distribution. □

Example 3.4. Let $\beta > 0$. Determine the pdf of the limiting case of the GB2 distribution when $\theta \to 0$, $\tau \to \infty$, and $\theta\tau^{\frac{1}{\gamma}} \to \beta$.

Solution. We follow the same procedure of Example 3.3 to derive the pdf of the limiting distribution. To do that, we first rewrite the GB2 density function as follows:

$$
\begin{aligned}
f(x) &= \frac{\Gamma(\alpha+\tau)}{\Gamma(\alpha)\Gamma(\tau)} \frac{\gamma(x/\theta)^{\gamma\tau}}{x[1+(x/\theta)^\gamma]^{\alpha+\tau}} \\
&= \frac{\Gamma(\alpha+\tau)}{e^{-\alpha-\tau}(\alpha+\tau)^{\alpha+\tau-\frac{1}{2}}\sqrt{2\pi}} \cdot \frac{e^{-\tau}\tau^{\tau-\frac{1}{2}}\sqrt{2\pi}}{\Gamma(\tau)} \cdot \\
&\quad \frac{e^{-\alpha-\tau}(\alpha+\tau)^{\alpha+\tau-\frac{1}{2}}\sqrt{2\pi}}{e^{-\tau}\tau^{\tau-\frac{1}{2}}\sqrt{2\pi}} \cdot \frac{\gamma}{\Gamma(\alpha)} \cdot \frac{(x/\theta)^{\gamma\tau}}{x[1+(x/\theta)^\gamma]^{\alpha+\tau}} \\
&= \frac{\Gamma(\alpha+\tau)}{e^{-\alpha-\tau}(\alpha+\tau)^{\alpha+\tau-\frac{1}{2}}\sqrt{2\pi}} \cdot \frac{e^{-\tau}\tau^{\tau-\frac{1}{2}}\sqrt{2\pi}}{\Gamma(\tau)} \cdot \\
&\quad \left(1+\frac{\alpha}{\tau}\right)^{\tau-\frac{1}{2}} \cdot \frac{\gamma e^{-\alpha}}{\Gamma(\alpha)} \cdot \frac{(\alpha+\tau)^\alpha (x/\theta)^{\gamma\tau}}{x[1+(x/\theta)^\gamma]^{\alpha+\tau}} \\
&= \frac{\Gamma(\alpha+\tau)}{e^{-\alpha-\tau}(\alpha+\tau)^{\alpha+\tau-\frac{1}{2}}\sqrt{2\pi}} \cdot \frac{e^{-\tau}\tau^{\tau-\frac{1}{2}}\sqrt{2\pi}}{\Gamma(\tau)} \cdot \\
&\quad \left(1+\frac{\alpha}{\tau}\right)^{\tau-\frac{1}{2}} \cdot \frac{\gamma e^{-\alpha}}{\Gamma(\alpha)} \cdot \frac{(1+\alpha/\tau)^\alpha x^{-\gamma\alpha}(\theta^\gamma\tau)^\alpha}{x[1+x^{-\gamma}/\theta^{-\gamma}]^{\alpha+\tau}}.
\end{aligned}
$$

By Theorem A.6, we have

$$
\begin{aligned}
&\lim_{\theta\to 0,\tau\to\infty,\theta\tau^{\frac{1}{\gamma}}\to\beta} [1+x^{-\gamma}/\theta^{-\gamma}]^{\alpha+\tau} \\
&= \lim_{\theta\to 0,\tau\to\infty,\theta\tau^{\frac{1}{\gamma}}\to\beta} \left([1+x^{-\gamma}/\theta^{-\gamma}]^{\theta^{-\gamma}}\right)^{\theta^\gamma\alpha+\theta^\gamma\tau} \\
&= e^{x^{-\gamma}\beta^\gamma}.
\end{aligned}
$$

Similarly as in Example 3.3, we have

$$
\lim_{\theta\to 0,\tau\to\infty,\theta\tau^{\frac{1}{\gamma}}\to\beta} f(x) = \frac{\gamma(\beta/x)^{\gamma\alpha}e^{-(\beta/x)^{\gamma\alpha}}}{x\Gamma(\alpha)},
$$

which is the density function of the inverse transformed gamma distribution. □

***

Exercise 3.12. Let $X \sim \text{GB2}(\alpha, \theta, \gamma, \tau)$. Suppose that $\gamma\tau > 1$. Show that X has a unique mode, which is given by

$$
x_{mode} = \theta\left(\frac{\gamma\tau - 1}{\gamma\alpha + 1}\right)^{\frac{1}{\gamma}}.
$$

Exercise 3.13. Let $X \sim \text{GB2}(\alpha, \theta, \gamma, \tau)$ and $\lambda > 0$. Show that

$$X^{\frac{1}{\lambda}} \sim \text{GB2}\left(\alpha, \theta^{\frac{1}{\lambda}}, \gamma\lambda, \tau\right).$$

Exercise 3.14. Let $X \sim \text{GB2}(\alpha, \theta, \gamma, \tau)$. Determine the distribution of X^{-1}.

Exercise 3.15. The Pareto distribution is a special case of the GB2 distribution. Let X follow the Pareto distribution with parameters α and θ. The pdf of X is

$$f(x) = \frac{\alpha\theta^\alpha}{(x+\theta)^{\alpha+1}}, \quad x > 0.$$

Show that when $\alpha > 1$,

$$E[X] = \frac{\theta}{\alpha - 1},$$

and when $\alpha > 2$,

$$\text{Var}(X) = \frac{\alpha\theta^2}{(\alpha-1)^2(\alpha-2)}.$$

Exercise 3.16. Let $X \sim \text{Beta}(\alpha, \beta)$, i.e., X follows the Beta distribution with parameters α and β, where $\alpha > 0$ and $\beta > 0$ are shape parameters. The pdf of X is defined as

$$f(x) = \frac{x^{\alpha-1}(1-x)^{\beta-1}}{B(\alpha,\beta)}, \quad x \in (0,1),$$

where $B(\alpha, \beta)$ is the beta function. Show that

$$E[X] = \frac{\alpha}{\alpha+\beta}, \quad \text{Var}(X) = \frac{\alpha\beta}{(\alpha+\beta)^2(\alpha+\beta+1)}.$$

Exercise 3.17. Compute the following integral:

$$\int_0^1 t^6(1-t)^6 \, \mathrm{d}t.$$

Exercise 3.18. Compute the following integral:

$$\int_0^\infty x^6(1+x)^{-14} \, \mathrm{d}x.$$

Exercise 3.19. A healthcare insurer implements an incentive program to control hospitalization. Under the incentive program, the physicians will be paid a bonus $B = c\max(400 - X, 0)$, where X is the total hospital claim amount and c is a constant. After the incentive program is implemented, the total hospital claim amount follows a Pareto distribution with $\alpha = 2$ and $\theta = 300$. Suppose that $E[B] = 100$. Determine c.

Exercise 3.20. An insurance agent receives bonus according to the following formula:

$$B = 0.15 \max(480,000 - X, 0),$$

where X is the incurred losses of the policies sold by the agent. Suppose that the incurred losses follow a Pareto distribution with parameters $\alpha = 2$ and $\theta = 500{,}000$. Calculate the expected value of the agent's bonus.

---***

4

Aggregate Loss Models

The severity models introduced in the previous chapter focus on a single loss or claim. In this chapter, we introduce aggregate loss models, which consider multiple losses or claims. In particular, we introduce the collective risk model and the individual risk model. In the collective risk model, the number of claims is a random variable; while in the individual risk model, the number of claims is a fixed number.

4.1 The Collective Risk Model

The collective risk model is presented in Definition 4.1. In the collective risk model, the number of claims is assumed to be a random variable. Given the number of claims, the individual losses are assumed to be independent and identically distributed. The collective risk model is suitable when the number of claims is unknown but the claim amounts have a common distribution.

Definition 4.1 (Collective risk model)**.** The collective risk model has the following form:

$$S = X_1 + X_2 + \cdots + X_N,$$

where N is the number of claims, X_i is the loss amount from the ith claim, and S denote the aggregate loss amount. The collective risk model imposes the following assumptions:

(a) Given $N = n$, the random variables X_1, X_2, ..., X_n are i.i.d. and the common distribution does not depend on n.

(b) The random variable N does not depend on X_1, X_2, ...

A major advantage of the collective risk model is that it allows the frequency (i.e., the number of claims) and the severity (i.e., the individual claim amount) to be modeled separately. The frequency can be modeled by frequency distributions introduced in Chapter 2. The severity can be modeled

DOI: 10.1201/9781003484899-4

by severity distributions introduced in Chapter 3. Once the frequency model and the severity model are determined, the distribution of the aggregate loss can be obtained from Theorem 4.1.

Theorem 4.1. *Let $\{p_k\}_{k\geq 0}$ be the pf of the frequency distribution. Let $F_X(x)$ be the cdf of the severity distribution with a positive support. Then the cdf of the aggregate loss is given by*

$$F_S(x) = \sum_{k=0}^{\infty} p_k F_X^{*k}(x), \quad x > 0, \tag{4.1}$$

where $F_X^{*k}(x)$ is the k-fold convolution of $F_X(x)$, i.e.,

$$F_X^{*k}(x) = \int_0^x F_X^{*(k-1)}(x-y) f_X(y)\,\mathrm{d}\,y, \quad F_X^{*0}(x) = \begin{cases} 0, & \text{if } x < 0, \\ 1, & \text{if } x \geq 0. \end{cases} \tag{4.2}$$

The pdf of the aggregate loss is given by

$$f_S(x) = \sum_{k=0}^{\infty} p_k f_X^{*k}(x), \quad x > 0, \tag{4.3}$$

where $f_X^{*k}(x)$ is the k-fold convolution of $f_X(x)$, i.e.,

$$f_X^{*k}(x) = \int_0^x f_X^{*(k-1)}(x-y) f_X(y)\,\mathrm{d}\,y, \quad f_X^{*0}(x) = \begin{cases} 1, & \text{if } x = 0, \\ 0, & \text{if } x \neq 0. \end{cases} \tag{4.4}$$

Proof. Since the event $\{S \leq x\}$ can be divided into the mutually exclusive events $\{S \leq x, N = 0\}$, $\{S \leq x, N = 1\}$, ..., we have

$$\begin{aligned} F_S(x) =& P(S \leq x) = \sum_{k=0}^{\infty} P(S \leq x, N = k) = \sum_{k=0}^{\infty} P(S \leq x | N = k) p_k \\ =& \sum_{k=0}^{\infty} P(X_1 + X_2 + \cdots + X_k \leq x) p_k. \end{aligned} \tag{4.5}$$

The probability $P(X_1 + X_2 + \cdots + X_k \leq x)$ can be calculated recursively. When $k = 0$, $S = 0$. Hence

$$P(0 \leq x) = \begin{cases} 0, & \text{if } x < 0, \\ 1, & \text{if } x \geq 0. \end{cases}$$

We let $F_X^{*0}(x) = P(0 \leq x)$ and

$$f_X^{*0}(x) = \begin{cases} 1, & \text{if } x = 0, \\ 0, & \text{if } x \neq 0. \end{cases}$$

When $k = 1$, $S = X_1$. In this case, $P(S \leq x) = P(X_1 \leq x) = F_X(x)$. We let $F_X^{*1}(x) = F_X(x)$ and $f_X^{*1}(x) = f_X(x)$. We continue this process and let

$$F_X^{*(k-1)}(x) = P(X_1 + X_2 + \cdots + X_{k-1} \leq x),$$

$$f_X^{*(k-1)}(x) = \frac{\mathrm{d}\, F_X^{*(k-1)}(x)}{\mathrm{d}\, x}.$$

Then

$$\begin{aligned}
&P(X_1 + X_2 + \cdots + X_k \leq x)\\
=&E\left[I_{\{X_1+\cdots+X_k\leq x\}}\right] = E\left[E\left[I_{\{X_1+\cdots+X_k\leq x\}}|X_k\right]\right]\\
=&\int_0^\infty E\left[I_{\{X_1+\cdots+X_k\leq x\}}|X_k = y\right] f_X(y)\,\mathrm{d}\,y\\
=&\int_0^\infty P(X_1 + X_2 + \cdots + X_{k-1} \leq x - y) f_X(y)\,\mathrm{d}\,y\\
=&\int_0^y F_X^{*(k-1)}(x-y) f_X(y)\,\mathrm{d}\,y.
\end{aligned}$$

Equation (4.1) follows by Equation (4.5) and letting

$$F_X^{*k}(x) = \int_0^y F_X^{*(k-1)}(x-y) f_X(y)\,\mathrm{d}\,y.$$

Equation (4.3) follows by Equation (4.5) and differentiating the above equation

$$f_X^{*k}(x) = \frac{\mathrm{d}\, F_X^{*k}(x)}{\mathrm{d}\, x} = \int_0^y f_X^{*(k-1)}(x-y) f_X(y)\,\mathrm{d}\,y.$$

□

In Theorem 4.1, we assume the common severity distribution has a positive support. The k-fold convolution of a distribution can be complicated. Example 4.1 shows that even the 2-fold convolution of the simple uniform distribution has a complex form.

Example 4.1. Let X and Y be uniform random variables on $[0, 1]$. Find the cdf and the pdf of $S = X + Y$.

Solution. The cdf and the pdf of the common distribution of X and Y are

$$F_X(x) = \begin{cases} 0, & \text{if } x < 0,\\ x, & \text{if } 0 \leq x \leq 1,\\ 1, & \text{if } 1 < x, \end{cases} \qquad f_X(x) = \begin{cases} 0, & \text{if } x < 0,\\ 1, & \text{if } 0 \leq x \leq 1,\\ 0, & \text{if } 1 < x, \end{cases}$$

The cdf of S is the 2-fold convolution of $F_X(x)$:

$$F_S(x) = \int_0^x F_X(x-y) f_X(y)\,\mathrm{d}\,y$$
$$= \begin{cases} \int_0^x (x-y)\,\mathrm{d}\,y, & \text{if } 0 \le x \le 1, \\ \int_0^{x-1} 1\,\mathrm{d}\,y + \int_{x-1}^1 (x-y)\,\mathrm{d}\,y, & \text{if } 1 < x \le 2 \end{cases}$$
$$= \begin{cases} \frac{1}{2}x^2, & \text{if } 0 \le x \le 1, \\ -\frac{1}{2}x^2 + 2x - 1, & \text{if } 1 < x \le 2. \end{cases}$$

The pdf of S is the 2-fold convolution of $f_X(x)$:

$$f_S(x) = \int_0^x f_X(x-y) f_X(y)\,\mathrm{d}\,y = \begin{cases} \int_0^x 1\,\mathrm{d}\,y, & \text{if } 0 \le x \le 1, \\ \int_{x-1}^1 1\,\mathrm{d}\,y, & \text{if } 1 < x \le 2 \end{cases}$$
$$= \begin{cases} x, & \text{if } 0 \le x \le 1, \\ 2 - x, & \text{if } 1 < x \le 2. \end{cases}$$

□

Theorem 4.2. *The mean and the variance of the collective risk model are given by*

$$E[S] = E[N]E[X],$$
$$\mathrm{Var}(S) = E[N]\,\mathrm{Var}(X) + \mathrm{Var}(N)E[X]^2.$$

Here $E[X]^2 = E[X] \times E[X] = (E[X])^2$.

Proof. Since N and X_i's are independent, the mean of S can be calculated as follows:

$$E[S] = E[E[S|N]] = E[NE[X]] = E[N]E[X].$$

Similarly, the second raw moment of S is

$$\begin{aligned} E[S^2] =& E[E[S^2|N]] = E[NE[X^2] + N(N-1)E[X]^2] \\ =& E[N]E[X^2] + E[N(N-1)]E[X]^2 \\ =& E[N]\,\mathrm{Var}(X) + E[N^2]E[X]^2. \end{aligned}$$

Hence the variance is

$$\begin{aligned} \mathrm{Var}(S) =& E[S^2] - E[S]^2 = E[N]\,\mathrm{Var}(X) + E[N^2]E[X]^2 - E[N]^2E[X]^2 \\ =& E[N]\,\mathrm{Var}(X) + \mathrm{Var}(N)E[X]^2. \end{aligned}$$

□

Example 4.2. In the collective risk model, the claim frequency follows a Poisson distribution with parameter $\lambda = 2$ and the claim severity follows a gamma distribution with parameters $\alpha = 3$ and $\theta = 1000$. Calculate $E[S]$ and $\text{Var}(S)$.

Solution. For the Poisson distribution, we have $E[N] = \text{Var}(N) = \lambda = 2$. For the gamma distribution, we have $E[X] = \alpha\theta = 3000$ and $\text{Var}(X) = \alpha\theta^2 = 3000000$. By Theorem 4.2, we have

$$E[S] = 2 \times 3000 = 6000,$$

and

$$\text{Var}(S) = 2 \times 3 \times 10^6 + 2 \times 3000^2 = 2.4 \times 10^7.$$

□

***

Exercise 4.1. Consider the collective risk model $S = X_1 + X_2 + \cdots + X_N$. Let $M_X(z)$ be the moment generating function of the claim severity. Let $P_N(t)$ be the probability generating function of the claim frequency. Show that the moment generating function of S is given by

$$M_S(z) = P_N(M_X(z)).$$

Exercise 4.2. Consider the collective risk model $S = X_1 + X_2 + \cdots + X_N$. Calculate $E[S]$ and $\text{Var}(S)$ by using the moment generating function of S that is given in Exercise 4.1.

Exercise 4.3. Let X_1 and X_2 be independent and identically distributed random variables with the following density function:

$$F_X(x) = \frac{x}{x+\theta}, \quad x > 0,$$

where $\theta > 0$. Find the cdf of $S = X_1 + X_2$.

Exercise 4.4. Let X_1, X_2, ..., and X_n be i.i.d random variables following the geometric distribution with parameter p, i.e.,

$$f_X(k) = p(1-p)^k, \quad k = 0, 1, 2, \ldots.$$

Find the pdf of $S = X_1 + X_2 + \cdots + X_n$.

Exercise 4.5. In the collective risk model, the number of claims N follows a Poisson distribution with a mean of 3 and the claim amount has the following distribution:

$$f_X(1) = \frac{1}{3}, \quad f_X(2) = \frac{2}{3}.$$

Calculate $f_S(4)$.

Exercise 4.6. In a collective risk model $S = X_1 + X_2 + \cdots + X_N$, the number of claims N follows a Poisson distribution with parameter λ. Show that the variance of the aggregate loss is

$$\mathrm{Var}(S) = E[N]E\left[X^2\right].$$

4.2 The Individual Risk Model

The individual risk model is presented in Definition 4.2. Unlike the collective risk model, the individual risk model assumes a fixed number of losses and does not assume that the individual loss amounts are identically distributed. The individual risk model is suitable to model the aggregate loss from a fixed number of insurance policies where the claim amounts from different policies may have different distributions.

Definition 4.2 (Individual risk model)**.** The individual risk model has the following form:

$$S = X_1 + X_2 + \cdots + X_n,$$

where n is a fixed number, X_i is the loss amount from the ith claim, and S is the aggregate loss amount. The random variables X_1, X_2, ..., X_n are assumed to be independent.

Example 4.3. Consider a portfolio of n life insurance policies. The total death benefit is $S = X_1 + X_2 + \cdots + X_n$, where X_i is the death benefit for the ith policyholder. For $i = 1, 2, \ldots, n$, X_i has the following distribution:

$$P(X_i = b_i) = q_i, \quad P(X_i = 0) = 1 - q_i,$$

where q_i is the probability of death. Calculate $E[S]$ and $\mathrm{Var}(S)$.

Solution. The mean of S can be calculated directly as follows:

$$E[S] = E\left[\sum_{i=1}^{n} X_i\right] = \sum_{i=1}^{n} E[X_i] = \sum_{i=1}^{n} b_i q_i.$$

By Theorem 1.6, we have

$$\mathrm{Var}(S) = \sum_{i=1}^{n} \mathrm{Var}(X_i) = \sum_{i=1}^{n} \left(b_i^2 q_i - b_i^2 q_i^2\right) = \sum_{i=1}^{n} b_i^2 q_i (1 - q_i).$$

□

The distribution of the sum of random variables is the convolution of the distributions of the random variables. A simple example is given in Example 4.4. However, the convolution of different individual distributions is complicated. As a result, obtaining explicit formulas for the probability distribution of the individual risk model is difficult. The distribution of S is usually approximated by other parametric distributions such as normal, gamma, and log-normal distributions.

Example 4.4. Let X be a uniform random variable on $[0, 1]$ and Y a uniform random variable on $[0, 2]$. Suppose that X and Y are independent. Find the pdf of $S = X + Y$.

Solution. The support of S is $[0, 3]$. The pdfs of X and Y are

$$f_X(x) = \begin{cases} 1, & \text{if } 0 \le x \le 1, \\ 0, & \text{otherwise,} \end{cases} \qquad f_Y(x) = \begin{cases} \frac{1}{2}, & \text{if } 0 \le x \le 2, \\ 0, & \text{otherwise.} \end{cases}$$

The pdf of S is the convolution of $f_X(x)$ and $f_Y(x)$. That is,

$$f_S(x) = \int_0^x f_X(x-y) f_Y(y)\,\mathrm{d}\,y = \begin{cases} \int_0^x 1 \cdot \frac{1}{2}\,\mathrm{d}\,y, & \text{if } 0 \le x \le 1, \\ \int_{x-1}^x 1 \cdot \frac{1}{2}\,\mathrm{d}\,y, & \text{if } 1 < x \le 2, \\ \int_{x-1}^2 1 \cdot \frac{1}{2}\,\mathrm{d}\,y, & \text{if } 2 < x \le 3, \\ 0, & \text{otherwise,} \end{cases}$$

$$= \begin{cases} \frac{x}{2}, & \text{if } 0 \le x \le 1, \\ \frac{1}{2}, & \text{if } 1 < x \le 2, \\ \frac{3-x}{2}, & \text{if } 2 < x \le 3, \\ 0, & \text{otherwise.} \end{cases}$$

□

Example 4.5. A small company purchased a group life insurance contract for its permanent employees. An actuary developed an individual risk model for the aggregate loss for the contract. Based on this individual risk model, the mean and the variance of the aggregate loss are given by:

$$E[S] = 2000, \quad \mathrm{Var}(S) = 10^8.$$

Approximate the probability that the aggregate loss are greater than 3000 by using the lognormal distribution.

Solution. The mean and the variance of the lognormal distribution with parameters μ and σ^2 are given by

$$\exp\left(\mu + \frac{1}{2}\sigma^2\right)$$

and

$$\exp\left(2\mu + \sigma^2\right)\left(\exp(\sigma^2) - 1\right),$$

respectively.

Suppose that the aggregate loss random variable S is approximately lognormal distributed with parameters (μ, σ^2). Then we can determine the parameters by

$$\exp\left(\mu + \frac{1}{2}\sigma^2\right) = E[S] = 2000$$
$$\exp\left(2\mu + \sigma^2\right)\left(\exp(\sigma^2) - 1\right) = \text{Var}(S) = 10^8.$$

Solving the above equations gives

$$\hat{\mu} = 5.9719, \quad \hat{\sigma} = 1.8050.$$

The probability can be approximated as follows:

$$\begin{aligned} P(S > 3000) =& P(\ln S > \ln 3000) = P\left(\frac{\ln S - \hat{\mu}}{\hat{\sigma}} > \frac{\ln 3000 - \hat{\mu}}{\hat{\sigma}}\right) \\ =& P\left(\frac{\ln S - \hat{\mu}}{\hat{\sigma}} > 1.13\right) \\ =& 1 - \Phi(1.13) = 1 - 0.8703 = 0.1297. \end{aligned}$$

□

***

Exercise 4.7. For $i = 1, 2$, let X_i follow the exponential distribution with parameter θ_i, i.e.,

$$f_{X_i}(x) = \frac{1}{\theta_i}\exp\left(-\frac{x}{\theta_i}\right), \quad x > 0.$$

Suppose that $\theta_1 \neq \theta_2$. Find the pdf of $S = X_1 + X_2$.

Exercise 4.8. Let X be an exponential random variable with parameter θ. Let Y be a uniform random variable on $[0, 1]$. Suppose that X and Y are independent. Find the pdf of $S = X + Y$.

Exercise 4.9. Let $n > 0$ be an integer. For $i = 1, 2, \ldots, n$, let X_i follow the gamma distribution with parameters $\alpha > 0$ and $\theta_i > 0$. Let $S = X_1 + X_2 + \cdots + X_n$. Suppose that X_1, X_2, $\ldots$, X_n are independent. Calculate $E[S]$ and $\text{Var}(S)$.

Exercise 4.10. Let S be a random variable with

$$E[S] = 2000, \quad \mathrm{Var}(S) = 10^8.$$

Approximate the probability $P(S > 3000)$ by using the normal distribution.

---***

5

Coverage Modifications

In this chapter, we introduce how to incorporate coverage modifications in loss modeling. In particular, we introduce deductibles, policy limits, coinsurance and stop-loss insurance.

5.1 Deductibles

Ordinary deductibles are the most common type of deductible. It is the amount taken out of the insurance payment. If the loss is less than the deductible, then the insurance company does not need to pay and there is no loss to the insurance company. There are two ways to model an ordinary deductible: on a per-payment basis and a per-loss basis. This is summarized in Definition 5.1.

Definition 5.1 (Ordinary deductible). An ordinary deductible is the amount of loss paid by a policyholder before an insurance policy will pay. Let d the the ordinary deductible amount and let X be the loss random variable. If an ordinary deductible is applied on a per-payment basis, the modified loss variable is

$$Y^P = \begin{cases} \text{undefined}, & \text{if } X \leq d, \\ X - d, & \text{if } X > d. \end{cases} \tag{5.1}$$

If an ordinary deductible is applied on a per-loss basis, the modified loss variable is

$$Y^L = (X - d)_+ = \begin{cases} 0, & \text{if } X \leq d, \\ X - d, & \text{if } X > d. \end{cases} \tag{5.2}$$

The random variable Y^P defined in Equation (5.1) is called the left truncated and shifted variable, or the per-payment variable, or the excess loss variable. The random variable Y^L defined in Equation (5.2) is called the left censored and shifted variable or the per-loss variable.

DOI: 10.1201/9781003484899-5

Example 5.1. Let X be a loss random variable that follows the exponential distribution with parameter $\theta = 1000$. Let Y^P and Y^L be the left truncated and shifted variable and the left censored and shifted variable for an ordinary deductible of 500, respectively. Determine the cdfs and the pdfs of Y^P and Y^L.

Solution. The cdf and the pdf of X are

$$F_X(x) = 1 - \exp\left(-\frac{x}{1000}\right), \quad f_X(x) = \frac{1}{1000}\exp\left(-\frac{x}{1000}\right), \quad x > 0.$$

The cdf of Y^P is given by

$$\begin{aligned} F_{Y^P}(x) =& P(Y^P \leq x) = P(X - d \leq x | X > d) = \frac{P(X - d \leq x, X > d)}{P(X > d)} \\ =& \frac{P(d < X \leq x + d)}{P(X > d)} = \frac{F_X(x+d) - F(d)}{1 - F(d)} \\ =& \frac{\exp\left(-\frac{d}{1000}\right) - \exp\left(-\frac{x+d}{1000}\right)}{\exp\left(-\frac{d}{1000}\right)} = 1 - \exp\left(-\frac{x}{1000}\right). \end{aligned}$$

The pdf of Y^P can be obtained by differentiating the cdf:

$$f_{Y^P}(x) = \frac{1}{1000}\exp\left(-\frac{x}{1000}\right), \quad x > 0.$$

The cdf of Y^L is given by

$$\begin{aligned} F_{Y^L}(x) =& P(Y^L \leq x) = P((X - d)_+ \leq x) = \begin{cases} P(X - d \leq 0), & \text{if } x = 0, \\ P(X - d \leq x), & \text{if } x > 0 \end{cases} \\ =& \begin{cases} F_X(d), & \text{if } x = 0, \\ F_X(x + d), & \text{if } x > 0 \end{cases} = \begin{cases} 1 - \exp\left(-\frac{1}{2}\right), & \text{if } x = 0, \\ 1 - \exp\left(-\frac{x + 500}{1000}\right), & \text{if } x > 0. \end{cases} \end{aligned}$$

The per-loss variable Y^L is a mixed random variable. Its pdf is given by

$$f_{Y^L}(x) = \begin{cases} 1 - \exp\left(-\frac{1}{2}\right), & \text{if } x = 0, \\ \frac{1}{1000}\exp\left(-\frac{x+500}{1000}\right), & \text{if } x > 0. \end{cases}$$

□

Example 5.1 shows how to find the cdfs and the pdfs of the per-payment variable and the per-loss variable for an ordinary deductible. It is interesting that the distribution of the per-payment variable is the same as that of the original loss variable. This is due to the memoryless property of the exponential distribution (see Exercise 3.7).

The expectation of the per-payment variable is called the mean excess loss. It is the conditional expectation of $X - d$ given $X > d$. See Definition 5.2. The expectation of the per-loss variable is related to the expectation of the per-payment variable proportionally (see Exercise 5.4).

Definition 5.2 (Mean excess loss). Let d be an ordinary deductible. Let X be a loss random variable such that $P(X > d) > 0$. Then the mean excess loss for the ordinary deductible is defined as

$$e_X(d) = e(d) = E\left[Y^P\right] = E[X - d|X > d] = \frac{\int_d^\infty (x-d) f_X(x)\,\mathrm{d}\,x}{1 - F_X(d)}.$$

In general, the kth moment of the excess loss variable is defined as

$$e_X^k(d) = E[(X-d)^k|X > d] = \frac{\int_d^\infty (x-d)^k f_X(x)\,\mathrm{d}\,x}{1 - F_X(d)}.$$

Inflation is an important factor to consider when calculating the expected cost after an ordinary deductible. When the loss is subject to uniform inflation, the per-payment variable and the per-loss variable are given by

$$Y^P = \begin{cases} \text{undefined}, & \text{if } (1+r)X \le d, \\ (1+r)X, & \text{if } (1+r)X > d, \end{cases}$$

and

$$Y^L = \begin{cases} 0, & \text{if } (1+r)X \le d, \\ (1+r)X, & \text{if } (1+r)X > d, \end{cases}$$

respectively, Theorem 5.1 shows how to calculate the expected cost after an ordinary deductible. The theorem also gives the expected cost when the loss is subject to uniform inflation.

Theorem 5.1. *Let X be a loss random variable and d an ordinary deductible. Let Y^P and Y^L be the resulting per-payment variable and the per-loss variable, respectively. Then the expected cost per-payment is*

$$E[Y^P] = e_X(d) = \frac{E[X] - E[X \wedge d]}{1 - F(d)}$$

and the expected cost per-loss is

$$E[Y^L] = E[(X-d)_+] = E[X] - E[X \wedge d],$$

where $X \wedge d = \min(X, d)$.

If the loss is subject to uniform inflation of $1+r$*, then the expected cost per-payment is*

$$E[Y^P] = \frac{1+r}{1-F\left(\dfrac{d}{1+r}\right)}\left(E[X] - E\left[X \wedge \frac{d}{1+r}\right]\right)$$

and the expected cost per-loss is

$$E[Y^L] = (1+r)\left(E[X] - E\left[X \wedge \frac{d}{1+r}\right]\right).$$

Proof. We only need to prove the part with inflation $1+r$. The part without inflation is the special case with $r = 0$. The expected cost per-payment can be calculated as follows:

$$\begin{aligned}
E[Y^P] =& \frac{\int_{d/(1+r)}^{\infty} ((1+r)x - d) f_X(x) \,\mathrm{d}\, x}{1 - F_X(d/(1+r))} \\
=& (1+r)\frac{\int_{d/(1+r)}^{\infty} (x - d/(1+r)) f_X(x) \,\mathrm{d}\, x}{1 - F_X(d/(1+r))} \\
=& (1+r)\frac{E[X] - \int_0^{d/(1+r)} x f_X(x) \,\mathrm{d}\, x - \int_{d/(1+r)}^{\infty} d/(1+r) f_X(x) \,\mathrm{d}\, x}{1 - F_X(d)} \\
=& (1+r)\frac{E[X] - \int_0^{\infty} \min(x, d/(1+r)) f_X(x) \,\mathrm{d}\, x}{1 - F_X(d/(1+r))} \\
=& (1+r)\frac{E[X] - E\left[X \wedge \dfrac{d}{1+r}\right]}{1 - F(d/(1+r))}.
\end{aligned}$$

The expected cost per-loss is obtained by multiplying the expected cost per-payment by $1 - F(d/(1+r))$. □

The loss elimination ratio is a quantity that can be used to evaluate the effect of an ordinary deductible. It is the ratio of the decrease in the expected payment with an ordinary deductible to the expected payment without the deductible. Mathematically, it is defined in Definition 5.3.

Definition 5.3 (Loss elimination ratio)**.** Let X be a loss random variable and d an ordinary deductible. Then the loss elimination ratio is defined as

$$\frac{E[X \wedge d]}{E[X]}.$$

If the loss is subject to uniform inflation of $1 + r$, the loss elimination ratio is

$$\frac{E\left[X \wedge \frac{d}{1+r}\right]}{E[X]}.$$

Example 5.2. Let X be a loss random variable that follows the Pareto distribution with parameters $\alpha = 2$ and $\theta = 1000$. The ordinary deductible is 500. Calculate the expected cost per loss and the loss elimination ratio without inflation. Then calculate the expected cost per loss and the loss elimination ratio with uniform inflation 10%.

Solution. Let us first calculate the expected cost per loss and the loss elimination ratio without inflation. By Theorem 5.1 and Exercise 5.3, we have

$$\begin{aligned} E[Y^L] =& E[X] - E[X \wedge d] = \frac{\theta}{\alpha - 1} - \frac{\theta}{\alpha - 1}\left[1 - \left(\frac{\theta}{d+\theta}\right)^{\alpha-1}\right] \\ =& \frac{\theta}{\alpha - 1}\left(\frac{\theta}{d+\theta}\right)^{\alpha-1} = \frac{1000}{1} \cdot \frac{1000}{500 + 1000} = 666.67. \end{aligned}$$

The loss elimination ratio is

$$\frac{E[X \wedge d]}{E[X]} = 1 - \left(\frac{\theta}{d+\theta}\right)^{\alpha-1} = 1 - \frac{1000}{500+1000} = 0.33.$$

The ordinary deductible eliminates 33% of the loss.

When the loss is subject to 10% inflation, the expected cost per loss is

$$\begin{aligned} E[Y^L] =& (1+r)\left(E[X] - E\left[X \wedge \frac{d}{1+r}\right]\right) \\ =& \frac{(1+r)\theta}{\alpha - 1}\left(\frac{\theta}{d/(1+r) + \theta}\right)^{\alpha-1} \\ =& \frac{1.1 \times 1000}{1} \cdot \frac{1000}{500/1.1 + 1000} = 756.25. \end{aligned}$$

The loss elimination ratio is

$$\frac{E\left[X \wedge \frac{d}{1+r}\right]}{E[X]} = 1 - \left(\frac{\theta}{d/(1+r) + \theta}\right)^{\alpha-1} = 1 - \frac{1000}{500/1.1 + 1000} = 0.3125.$$

□

A different type of deductible is the franchise deductible, which is defined in Definition 5.4. If the loss amount is greater than the franchise deductible, the insurance policy will pay the full loss amount.

Definition 5.4 (Franchise deductible)**.** A franchise deductible is the minimum amount of loss that must be incurred before an insurance policy pays. Let d the the deductible amount and let X be the loss random variable. If a franchise deductible is applied on a per-payment basis, the modified loss variable is

$$Y^P = \begin{cases} \text{undefined}, & \text{if } X \leq d, \\ X, & \text{if } X > d. \end{cases} \tag{5.3}$$

If a franchise deductible is applied on a per-loss basis, the modified loss variable is

$$Y^L = \begin{cases} 0, & \text{if } X \leq d, \\ X, & \text{if } X > d. \end{cases} \tag{5.4}$$

Example 5.3. Let X be a loss random variable that follows the exponential distribution with parameter $\theta = 1000$. Let Y^P and Y^L be the per-payment and per-loss variables for a franchise deductible of 500, respectively. Determine the cdfs and the pdfs of Y^P and Y^L.

Solution. The cdf of Y^P is given by

$$\begin{aligned} F_{Y^P}(x) =& P(Y^P \leq x) = P(X \leq x | X > d) = \frac{P(X \leq x, X > d)}{P(X > d)} \\ =& \frac{P(d < X \leq x)}{P(X > d)} = \begin{cases} 0, & \text{if } x \leq d, \\ \dfrac{F_X(x) - F(d)}{1 - F(d)}, & \text{if } x > d \end{cases} \\ =& \begin{cases} 0, & \text{if } x \leq 500, \\ 1 - \exp\left(-\dfrac{x - 500}{1000}\right), & \text{if } x > 500. \end{cases} \end{aligned}$$

The pdf of Y^P can be obtained by differentiating the cdf:

$$f_{Y^P}(x) = \frac{1}{1000} \exp\left(-\frac{x - 500}{1000}\right), \quad x > 500.$$

The cdf of Y^L is given by

$$\begin{aligned} F_{Y^L}(x) =& P(Y^L \leq x) = \begin{cases} P(X \leq d), & \text{if } x \leq d, \\ P(X \leq x), & \text{if } x > d \end{cases} \\ =& \begin{cases} F_X(d), & \text{if } x \leq d, \\ F_X(x), & \text{if } x > d \end{cases} = \begin{cases} 1 - \exp\left(-\dfrac{1}{2}\right), & \text{if } x \leq d, \\ 1 - \exp\left(-\dfrac{x}{1000}\right), & \text{if } x > d. \end{cases} \end{aligned}$$

The pdf of Y^L is given by

$$f_{Y^L}(x) = \begin{cases} 1 - \exp\left(-\frac{1}{2}\right), & \text{if } x = 0, \\ \frac{1}{1000}\exp\left(-\frac{x}{1000}\right), & \text{if } x > d. \end{cases}$$

□

***

Exercise 5.1. Let X be a loss random variable that follows the Pareto distribution with parameters $\alpha = 1$ and $\theta = 1000$, i.e.,

$$f_X(x) = \frac{1000}{(x+1000)^2}, \quad x > 0.$$

Let Y^P and Y^L be the left truncated and shifted variable and the left censored and shifted variable for an ordinary deductible of 500. Determine the cdfs and the pdfs of Y^P and Y^L.

Exercise 5.2. Let X be a continuous random variable. Let $S(x)$ be the survival function of X. Let d be a constant such that $S(d) > 0$. Show that

$$e_X(d) = \frac{\int_d^\infty S(x)\,\mathrm{d}x}{S(d)}.$$

Exercise 5.3. Let X be a random variable following the Pareto distribution with parameters $\alpha > 1$ and $\theta > 0$, i.e.,

$$f(x) = \frac{\alpha\theta^\alpha}{(x+\theta)^{\alpha+1}}, \quad x > 0.$$

Show that

$$E[X \wedge d] = \frac{\theta}{\alpha - 1}\left[1 - \left(\frac{\theta}{d+\theta}\right)^{\alpha-1}\right]$$

and

$$e_X(d) = \frac{d+\theta}{\alpha-1}.$$

Exercise 5.4. Let X be a continuous random variable. Let d be a constant such that $F(d) < 1$. Show that for $k \geq 1$,

$$E\left[(X-d)_+^k\right] = e_X^k(d)[1 - F(d)].$$

Exercise 5.5. Let X be a loss random variable and d a franchise deductible. Let Y^P and Y^L be the per-payment variable and the per-loss variable for the franchise deductible, respectively. Show that

$$E[Y^P] = \frac{E[X] - E[X \wedge d]}{1 - F(d)} + d$$

and

$$E[Y^L] = E[X] - E[X \wedge d] + d(1 - F(d)).$$

Exercise 5.6. Let X be a loss random variable with the following pdf:

$$f_X(x) = \begin{cases} 0.01, & \text{if } x \leq 80, \\ \dfrac{120 - x}{4000}, & \text{if } 80 < x \leq 120. \end{cases}$$

Calculate the loss elimination ratio for an ordinary deductible of 20.

Exercise 5.7. The losses from a portfolio of policies follow the Pareto distribution with parameters $\alpha = 2$ and $\theta = 5$. Calculate the loss elimination ratio for an ordinary deductible of 10 with a 20% uniform inflation rate.

Exercise 5.8. Losses from a portfolio of policies follow the Pareto distribution with parameters $\alpha = 2$ and $\theta = 1000$. The loss elimination ratio for an ordinary deductible is 70%. Determine the ordinary deductible.

---***

5.2 Policy Limits

Policy limits are a common feature of insurance policies. Definition 5.5 gives the definition of the policy limit and the modified loss variable.

Definition 5.5 (Policy limit)**.** A policy limit is the maximum amount of loss that an insurance company will pay. Let u be the policy limit and X the loss random variable. The modified loss variable is

$$Y = \begin{cases} X, & \text{if } X \leq u, \\ u, & \text{if } X > u. \end{cases}$$

The modified loss variable for a policy limit is a right-censored random variable and has a mixed distribution. In particular, its cdf and pdf are given by

$$F_Y(x) = \begin{cases} F_X(x), & \text{if } x < u, \\ 1, & \text{if } x \geq u, \end{cases}$$

and

$$f_Y(x) = \begin{cases} f_X(x), & \text{if } x < u, \\ 1 - F_X(u), & \text{if } x = u, \end{cases}$$

respectively. Like ordinary deductibles, policy limits are also affected by inflation. After uniform inflation $1 + r$, the modified loss variable is

$$Y = \begin{cases} (1+r)X, & \text{if } (1+r)X \leq u, \\ (1+r)u, & \text{if } (1+r)X > u. \end{cases}$$

Theorem 5.2. *Let X be a loss random variable and u a policy limit. Let r be the uniform inflation rate. Then the expected cost is*

$$(1+r)E\left[X \wedge \frac{u}{1+r}\right].$$

Proof. When the loss is subject to uniform inflation, the loss random variable becomes $(1+r)X$. By Exercise 5.9, we have

$$E[(1+r)X \wedge u] = (1+r)E\left[X \wedge \frac{u}{1+r}\right].$$

□

Example 5.4. Let X be a loss random variable that follows the Pareto distribution with parameters $\alpha = 2$ and $\theta = 1000$. A policy limit of 3000 is imposed. Calculate the expected cost without inflation and with uniform inflation 10%.

Solution. By Theorem 5.2, the expected cost without inflation is

$$\begin{aligned} E[Y] =& E[X \wedge u] = \frac{\theta}{\alpha-1}\left[1 - \left(\frac{\theta}{u+\theta}\right)^{\alpha-1}\right] \\ =& \frac{1000}{1} \cdot \left(1 - \frac{1000}{3000+1000}\right) = 750. \end{aligned}$$

The expected cost with 10% uniform inflation is

$$\begin{aligned} E[Y] =& (1+r)E\left[X \wedge \frac{u}{1+r}\right] = \frac{(1+r)\theta}{\alpha-1}\left[1 - \left(\frac{\theta}{u/(1+r)+\theta}\right)^{\alpha-1}\right] \\ =& \frac{1.1 \times 1000}{1} \cdot \left(1 - \frac{1000}{3000/1.1+1000}\right) = 804.88. \end{aligned}$$

With 10% uniform inflation, the expected cost increases

$$\frac{804.88 - 705}{750} = 7.32\%,$$

which is less than the inflation rate. □

Example 5.4 illustrates how to calculate the expected cost after a policy limit is imposed. Comparing this example with Example 5.2, we see that the impact of inflation on ordinary deductibles is different from that on policy limits. After applying 10% uniform inflation, the expected cost per loss for an ordinary deductible increases by more than 10%. However, the expected cost for a policy limit increases by less than 10%.

***

Exercise 5.9. Let X be a random variable. Let b and $c > 0$ be constants. Show that

$$E[cX \wedge b] = cE\left[X \wedge \frac{b}{c}\right].$$

Exercise 5.10. Let X be an exponential random variable with parameter $\theta > 0$, i.e.,

$$f_X(x) = \frac{1}{\theta}\exp\left(-\frac{x}{\theta}\right), \quad x > 0.$$

Show that for $u > 0$,

$$E[X \wedge u] = \theta\left(1 - \exp\left(-\frac{u}{\theta}\right)\right).$$

Exercise 5.11. Let X be a loss random variable that follows the exponential distribution with parameter $\theta = 1000$. A policy limit of 3000 is imposed. Calculate the expected cost without inflation and with uniform inflation 10%.

Exercise 5.12. The auto claim amount follows a distribution with the following cdf:

$$F(x) = 1 - 0.8e^{-0.02x} - 0.2e^{-0.001x}, \quad x \geq 0.$$

An insurance policy pays amounts up to a limit of 1000 per claim. Calculate the expected payment under this policy for one claim.

***

5.3 Coinsurance

Deductibles and policy limits help reduce the expected cost of the insurance company. With deductibles and policy limits, the policyholders also pay less premiums. Coinsurance is another coverage modification to reduce the expected cost and the premium.

Definition 5.6 (Coinsurance). Coinsurance is a coverage modification in which the insurance company pays a proportion of the loss and the policyholder pays the remaining loss. Let X be the loss random variable and c the proportion paid by insurance. The modified loss variable is

$$Y = cX.$$

Definition 5.6 gives the modified loss variable after coinsurance is applied. The modified loss variable is a linear transformation of the original loss variable (see Definition 1.12). An insurance policy may include multiple modifications. When a policy has an ordinary deductible, a policy limit, and coinsurance, we can derive the per-loss variable. Definition 5.7 shows the modified loss variable when all modifications introduced so far are present.

Definition 5.7 (Per-loss variable with multiple modifications)**.** Let X be a loss random variable. Let d be an ordinary deductible, u a policy limit, c the proportion paid by insurance, and r the uniform inflation rate. When all modifications are present, the per-loss variable is

$$Y^L = \begin{cases} 0, & \text{if } X < \dfrac{d}{1+r}, \\ c\left[(1+r)X - d\right], & \text{if } \dfrac{d}{1+r} \le X < \dfrac{u}{1+r}, \\ c(u-d), & \text{if } X \ge \dfrac{u}{1+r}. \end{cases}$$

The per-payment variable is

$$Y^P = \begin{cases} \text{undefined}, & \text{if } X < \dfrac{d}{1+r}, \\ c\left[(1+r)X - d\right], & \text{if } \dfrac{d}{1+r} \le X < \dfrac{u}{1+r}, \\ c(u-d), & \text{if } X \ge \dfrac{u}{1+r}. \end{cases}$$

Theorem 5.3 gives the expected cost of the per-loss variable and the per-payment variable in the general case when all coverage modifications are present. Theorems 5.1 and 5.2 are special cases of Theorem 5.3.

Theorem 5.3. *Let X be a loss random variable. Let d be an ordinary deductible, u a policy limit, c the proportion paid by insurance, and r the uniform inflation rate. Then the mean of the per-loss variable is given by*

$$E[Y^L] = c(1+r)\left(E\left[X \wedge \frac{u}{1+r}\right] - E\left[X \wedge \frac{d}{1+r}\right]\right).$$

The second moment of the per-loss variable is given by

$$\begin{aligned} E\left[\left(Y^L\right)^2\right] =& c^2(1+r)^2\left(E\left[\left(X \wedge \frac{u}{1+r}\right)^2\right] - E\left[\left(X \wedge \frac{d}{1+r}\right)^2\right]\right) \\ &- 2c^2d(1+r)\left(E\left[X \wedge \frac{u}{1+r}\right] - E\left[X \wedge \frac{d}{1+r}\right]\right) \end{aligned}$$

The mean and the second raw moment of the per-payment variable can be obtained by dividing the mean and the variance of the per-loss variable by $1 - F_X(d/(1+r))$, respectively.

Proof. The per-loss variable Y^L given in Definition 5.7 can be expressed as

$$Y^L = c(1+r)\left(X \wedge \frac{u}{1+r} - X \wedge \frac{d}{1+r}\right).$$

Hence the mean of the per-loss variable is calculated as

$$\begin{aligned}
E[Y^L] =& E\left[c(1+r)\left(X \wedge \frac{u}{1+r} - X \wedge \frac{d}{1+r}\right)\right] \\
=& c(1+r)\left(E\left[X \wedge \frac{u}{1+r}\right] - E\left[X \wedge \frac{d}{1+r}\right]\right).
\end{aligned}$$

To calculate the second raw moment of Y^L, we first expand $\left(Y^L\right)^2$ as follows:

$$\begin{aligned}
\left(Y^L\right)^2 =& c^2(1+r)^2\left(X \wedge \frac{u}{1+r} - X \wedge \frac{d}{1+r}\right)^2 \\
=& c^2(1+r)^2\left(X \wedge \frac{u}{1+r}\right)^2 + c^2(1+r)^2\left(X \wedge \frac{d}{1+r}\right)^2 \\
&- 2c^2(1+r)^2\left(X \wedge \frac{u}{1+r}\right)\left(X \wedge \frac{d}{1+r}\right) \\
=& c^2(1+r)^2\left(X \wedge \frac{u}{1+r}\right)^2 - c^2(1+r)^2\left(X \wedge \frac{d}{1+r}\right)^2 \\
&- 2c^2(1+r)^2\left(X \wedge \frac{d}{1+r}\right)\left(X \wedge \frac{u}{1+r} - X \wedge \frac{d}{1+r}\right) \\
=& c^2(1+r)^2\left(X \wedge \frac{u}{1+r}\right)^2 - c^2(1+r)^2\left(X \wedge \frac{d}{1+r}\right)^2 \\
&- 2c^2(1+r)d\left(X \wedge \frac{u}{1+r} - X \wedge \frac{d}{1+r}\right).
\end{aligned}$$

The last step follows from Exercise 5.13. The second raw moment of Y^L follows immediately from the above equation.

The moments of the per-payment variable are the conditional moments of Y^L given $X > d/(1+r)$. These moments can be obtained by dividing the moments of Y^L by $1 - F(d/(1+r))$. □

Example 5.5. Let X be a loss random variable that follows the Pareto distribution with parameters $\alpha = 2$ and $\theta = 1000$. An ordinary deductible of 500 and a policy limit of 3000 are imposed. Calculate the expected cost per-loss without inflation and with uniform inflation 10%.

Solution. By Theorem 5.3, the expected cost per-loss without inflation is

$$\begin{aligned}
E[Y^L] =& E[X \wedge u] - E[X \wedge d] \\
=& \frac{\theta}{\alpha-1}\left[1-\left(\frac{\theta}{u+\theta}\right)^{\alpha-1}\right] - \frac{\theta}{\alpha-1}\left[1-\left(\frac{\theta}{d+\theta}\right)^{\alpha-1}\right] \\
=& \frac{\theta}{\alpha-1}\left[\left(\frac{\theta}{d+\theta}\right)^{\alpha-1} - \left(\frac{\theta}{u+\theta}\right)^{\alpha-1}\right] \\
=& \frac{1000}{1}\cdot\left(\frac{1000}{500+1000} - \frac{1000}{3000+1000}\right) = 416.67.
\end{aligned}$$

The expected cost per-loss with 10% uniform inflation is

$$\begin{aligned}
E[Y^L] =& (1+r)(E\left[X \wedge \frac{u}{1+r}\right] - E\left[X \wedge \frac{d}{1+r}\right]) \\
=& \frac{(1+r)\theta}{\alpha-1}\left[1-\left(\frac{\theta}{u/(1+r)+\theta}\right)^{\alpha-1}\right] \\
& - \frac{(1+r)\theta}{\alpha-1}\left[1-\left(\frac{\theta}{d/(1+r)+\theta}\right)^{\alpha-1}\right] \\
=& \frac{(1+r)\theta}{\alpha-1}\left[\left(\frac{\theta}{d/(1+r)+\theta}\right)^{\alpha-1} - \left(\frac{\theta}{u/(1+r)+\theta}\right)^{\alpha-1}\right] \\
=& \frac{1.1\times 1000}{1}\cdot\left(\frac{1000}{500/1.1+1000} - \frac{1000}{3000/1.1+1000}\right) = 461.13.
\end{aligned}$$

□

***

Exercise 5.13. Let X be a random variable. Let d and u be constants such that $d < u$. Show that

$$(X \wedge d)(X \wedge u - X \wedge d) = d(X \wedge u - X \wedge d).$$

Exercise 5.14. A prescription drug plan has the following provisions:

(a) The ordinary deductible per year is 250.

(b) The policyholder pays 25% of the costs between 250 and 2250.

(c) The policyholder pays 100% of the costs above 2250 until the policyholder has paid 3600 in total.

(d) The policyholder pays 5% of the remaining costs.

Suppose that the costs follow a Pareto distribution with parameter $\alpha = 2$ and $\theta = 1000$. Calculate the expected annual payment from the plan.

Exercise 5.15. A loss random variable X has the following characteristics:

$$E[X] = 70, \quad E[X \wedge 30] = 25, \quad P(X > 30) = 0.75, \quad E\left[X^2 | X > 30\right] = 9000.$$

Calculate the variance of the per-loss variable for an ordinary deductible of 30.

Exercise 5.16. An insurance plan has the following provisions:

(a) There is an ordinary deductible of 20 per loss.

(b) The plan pays 80% of the amount of the loss in excess of 20.

(c) The maximum amount the plan pays is 60.

Suppose that the losses has the following distribution:

$$f_X(x) = \frac{x}{5000}, \quad 0 \leq x \leq 100.$$

Calculate the mean of the per-payment variable.

Exercise 5.17. A loss random variable X has the following distribution:

$$P(X = 100) = 0.2, \quad P(X = 200) = 0.2, \quad P(X = 300) = 0.6.$$

Calculate the variance of the per-payment variable for an ordinary deductible of 150.

***

5.4 Stop-loss Insurance

Stop-loss insurance also involves a deductible. However, the deductible in stop-loss insurance applies to aggregate losses rather than individual losses.

Definition 5.8 (Stop-loss insurance). Stop-loss insurance is insurance on the aggregate losses that is subject to a deductible. Let S be the aggregate loss variable. The expected cost of stop-loss insurance with deductible d is given by

$$E[(S - d)_+].$$

The expected cost is also called the net stop-loss premium.

Given the aggregate loss distribution, the net stop-loss premium can be calculated as follows:

$$E[(S-d)_+] = \int_d^\infty (x-d) f_S(x)\, \mathrm{d}\, x = \int_d^\infty (1 - F_S(x))\, \mathrm{d}\, x. \tag{5.5}$$

When S does not take values in an interval, then the net stop-loss premium for a deductible in the interval can be interpolated from those for deductibles at the end points. This is summarized in Theorem 5.4.

Theorem 5.4. *Let S be an aggregate loss variable. Let $a < b$. Suppose that $P(a < S < b) = 0$. Then*

$$E[(S-c)_+] = \frac{b-c}{b-a} E[(S-a)_+] + \frac{c-a}{b-a} E[(S-b)_+]$$

for all $c \in [a, b]$.

Proof. Since $P(a < S < b) = 0$, we have for all $x \in [a, b]$,

$$F_S(x) = P(S \le x) = P(S \le a) + P(a < S \le x) = P(S \le a) = F_S(a).$$

By definition and the above equation, we have

$$\begin{aligned} E[(S-c)_+] = & \int_c^\infty (1 - F_S(x))\, \mathrm{d}\, x = \int_a^\infty (1 - F_S(x))\, \mathrm{d}\, x - \int_a^c (1 - F_S(x))\, \mathrm{d}\, x \\ = & E[(S-a)_+] - (c-a)(1 - F_S(a)) \end{aligned}$$

and

$$\begin{aligned} E[(S-c)_+] = & \int_c^\infty (1 - F_S(x))\, \mathrm{d}\, x = \int_b^\infty (1 - F_S(x))\, \mathrm{d}\, x + \int_c^b (1 - F_S(x))\, \mathrm{d}\, x \\ = & E[(S-b)_+] + (b-c)(1 - F_S(a)). \end{aligned}$$

The result follows from the above two equations by canceling the term $1 - F_S(a)$. □

Example 5.6. Let $S = X_1 + X_2 + X_3$, where X_1, X_2, and X_3 are i.i.d and the common distribution is

$$P(X=0) = 0.4, \quad P(X=1) = 0.3, \quad P(X=2) = 0.2, \quad P(X=3) = 0.1.$$

Calculate $E[(S-1)_+]$.

Solution. Since S is the aggregate loss from three individual losses and each individual loss variable has the support $\{0, 1, 2, 3\}$, the support of S

is $\{0, 1, \ldots, 9\}$. Let p_0, p_1, ..., p_9 be the probability function of S. The probability p_0 is

$$p_0 = P(S = 0) = P(X_1 = X_2 = X_3 = 0) = 0.4^3 = 0.064.$$

Then by definition, we have

$$\begin{aligned} E[(S-1)_+] &= \sum_{x=1}^{9}(x-1)p_x = \sum_{x=1}^{x} p_x - \sum_{x=1}^{9} p_x = E[S] - (1 - p_0) \\ &= 3E[X] - 1 + 0.064 = 3(1 \times 0.3 + 2 \times 0.2 + 3 \times 0.1) - 1 + 0.064 \\ &= 2.064. \end{aligned}$$

□

***

Exercise 5.18. Let S be an aggregate loss variable with support $\{0, h, 2h, \ldots\}$, where $h > 0$ is a constant. Show that for $j \geq 0$,

$$E[(S - (j+1)h)_+] = E[(S - jh)_+] - h(1 - F_S(jh)).$$

Exercise 5.19. Let $S = X_1 + X_2 + X_3$, where X_1, X_2, and X_3 are i.i.d and the common distribution is

$$P(X = 0) = 0.4, \quad P(X = 1) = 0.3, \quad P(X = 2) = 0.2, \quad P(X = 3) = 0.1.$$

Calculate $E[(S - 2)_+]$.

Exercise 5.20. Let $S = X_1 + X_2 + \cdots + X_N$, where N follows a Poisson distribution with parameter $\lambda = 2$. Given $N = n$, X_1, X_2, ..., X_n are i.i.d and the common distribution is

$$P(X = 1) = P(X = 2) = P(X = 3) = \frac{1}{3}.$$

Calculate $E[(S - 2)_+]$.

Exercise 5.21. You are given the following information about the aggregate claims S:

(a) Losses can only occur in multiples of 100.

(b) $F_S(200) = 0.5$, $F_S(300) = 0.65$, $F_S(400) = 0.75$.

(c) $E[(S - 300)_+] = 60$.

Calculate the net stop-loss premium for stop-loss insurance with deductible 325.

Exercise 5.22. Let S be an aggregate loss variable with the following properties:

(a) $E[(S-100)_+] = 15$.

(b) $E[(S-120)_+] = 10$.

(c) $P(80 < S < 120) = 0$.

Determine $P(S \le 80)$.

---***

5.5 Impact of Coverage Modifications on Frequency

Among the several coverage modifications introduced in previous sections, deductibles have a significant impact on the frequency distribution. When a deductible is imposed or increased, the number of payments will decrease. When a deductible is lowered, the number of payments will increase.

Let $X_1, X_2, \ldots, X_N$ be the loss amounts, where N is the number of claims. When an ordinary deductible is imposed, the number of claims will be

$$M = \sum_{i=1}^{N} I_{\{X_i > d\}}. \tag{5.6}$$

Suppose that given $N = n$, $X_1, X_2, \ldots, X_n$ are i.i.d. Then the mean of the new frequency variable can be calculated as follows:

$$\begin{aligned} E[M] =& E[E[M|N]] = E\left[\sum_{i=1}^{N} E[I_{\{X_i > d\}}]\right] \\ =& E\left[\sum_{i=1}^{N} P(X_i > d)\right] = E[NP(X_1 > d)] = P(X_1 > d)E[N]. \end{aligned} \tag{5.7}$$

From Equation (5.7), we see that the mean of the modified frequency variable after an ordinary deductible is imposed is a fraction of the original frequency variable.

Example 5.7. The losses from a portfolio of insurance policies follow the uniform distribution on $[0, b]$. The number of losses follows the Poisson distribution with parameter $\lambda > 0$. An ordinary deductible of $d \in (0, b)$ is imposed. Suppose that the deductible does not change the loss distribution. Calculate the mean of the modified frequency distribution.

Solution. Let X denote the loss random variable. Since the losses follow the uniform distribution on $[0, b]$, we have

$$P(X > d) = \int_d^b \frac{1}{b} \,\mathrm{d}x = \frac{b-d}{b}.$$

By Equation (5.7), we have

$$E[M] = P(X > d)E[N] = \frac{b-d}{b}\lambda.$$

□

***

Exercise 5.23. The claim severity of a group dental policy follow the following distribution:

$$P(X = 40) = P(X = 80) = P(X = 120) = P(X = 200) = \frac{1}{4}.$$

The number of claims N follows a negative binomial distribution with $E[N] = 300$ and $\mathrm{Var}(N) = 800$. The group dental policy will experience the following changes:

(a) The claim severity will increase 50%.

(b) An ordinary deductible of 100 per claim will be imposed.

Calculate the expected number of payments after those changes.

***

6

Model Estimation

We have introduced models for the number of losses, the severity of losses, and aggregate losses in previous chapters. In this chapter, we introduce methods to estimate parameters of those models.

6.1 The Method of Moments

The method of moments provides a simple way to estimate model parameters. The idea of this method is to match the raw moments of the model with the empirical estimates of the raw moments.

Definition 6.1 (Method of moments)**.** Let x_1, x_2, ..., x_n be n observations of a random variable X. Let $\boldsymbol{\theta} = (\theta_1, \theta_2, \ldots, \theta_p)$ be the vector of parameters of a probability model for X. A method of moments estimate of $\boldsymbol{\theta}$ is any solution of the following p equations:

$$E\left[X^k|\boldsymbol{\theta}\right] = \frac{1}{n}\sum_{i=1}^{n} x_i^k, \quad k = 1, 2, \ldots, p.$$

Suppose that a model has p parameters. Then we can estimate the p parameters by matching the first p raw moments. Definition 6.1 summarizes the method of moments. Example 6.1 illustrates how the method of moments is used to estimate parameters.

Example 6.1. Consider the following dataset:

k	0	1	2	3	4	5
n_k	20	25	30	15	8	2

Assume that the data follow a binomial distribution with parameters m and q. Estimate the parameters by using the method of moments.

DOI: 10.1201/9781003484899-6

Solution. The first raw moment and the second raw moment of the binomial distribution are given by (see Theorem 2.1):

$$E[N] = mq, \quad E[N^2] = mq(1-q) + m^2q^2.$$

The empirical estimates of the raw moments are

$$\begin{aligned}
\hat{\mu}_1' &= \frac{\sum_{k=0}^{5} kn_k}{\sum_{k=0}^{5} n_k} = \frac{25+60+45+32+10}{100} \\
&= 1.72, \\
\hat{\mu}_2' &= \frac{\sum_{k=0}^{5} k^2 n_k}{\sum_{k=0}^{5} n_k} = \frac{25+120+135+128+50}{100} \\
&= 4.58.
\end{aligned}$$

Matching the raw moments from the model and empirical estimates, we get

$$\begin{aligned}
mq &= 1.72, \\
mq(1-q) + m^2q^2 &= 4.58.
\end{aligned}$$

Solving the equations gives

$$\hat{m} = 30, \quad \hat{q} = 0.0572.$$

□

***

Exercise 6.1. The random variable X is uniformly distributed on $[a, b]$, where $a < b$. You are given the following five observations:

$$9, \quad 33, \quad 41, \quad 77, \quad 91.$$

Estimate a and b by the method of moments.

Exercise 6.2. A loss random variable X follows the exponential distribution with parameter θ. A sample of X is:

$$98, \quad 399, \quad 1067, \quad 1273, \quad 1366.$$

Estimate θ by the method of moments.

Exercise 6.3. A loss random variable X follows the Pareto distribution with parameter α and θ. A sample of X is:

$$100, \quad 235, \quad 281, \quad 937, \quad 5843.$$

Estimate α and θ by the method of moments.

***

6.2 Maximum Likelihood Estimation

The maximum likelihood method is a commonly used method to estimate parameters of probability distributions from observed data. In the maximum likelihood method, the parameters are estimated by maximizing the likelihood function. Definition 6.2 gives the definition of the likelihood function.

Definition 6.2 (Likelihood function)**.** Let $\boldsymbol{\theta}$ be the vector of parameters of a probability model. Let E_1, E_2, ..., E_n be a list of observed events. Suppose that these events are independent. Then the likelihood function is defined by

$$L(\boldsymbol{\theta}) = \prod_{i=1}^{n} P(E_i|\boldsymbol{\theta}).$$

The log-likelihood function is given by

$$l(\boldsymbol{\theta}) = \sum_{i=1}^{n} \ln P(E_i|\boldsymbol{\theta}).$$

Maximizing the likelihood function is equivalent to maximizing the log-likelihood function. In most situations, it is easier to work with the log-likelihood function than the likelihood function. When n is large, the value of the likelihood function is a small number that is close to zero. For example, suppose that there are 100 observed events and the probability of each event is around 0.5. Then the value of the likelihood function will be

$$0.5^{100} = 7.888609 \times 10^{-31}.$$

The value of the corresponding log-likelihood function is

$$\ln\left(0.5^{100}\right) = 100\ln(0.5) = -69.31472.$$

Example 6.2. Losses follow the exponential distribution with parameter θ. A random sample of five losses was produced. The values of the three losses are 3, 6 and 14. The values of two losses are greater than 25. Calculate the maximum likelihood estimate of θ.

Solution. Let X be the loss random variable. Since X follows the exponential distribution with parameter θ, the probabilities of the five events are

$$P(X=3|\theta) = \frac{1}{\theta}\exp\left(-\frac{3}{\theta}\right), \quad P(X=6|\theta) = \frac{1}{\theta}\exp\left(-\frac{6}{\theta}\right),$$
$$P(X=14|\theta) = \frac{1}{\theta}\exp\left(-\frac{14}{\theta}\right), \quad P(X>25|\theta) = \exp\left(-\frac{25}{\theta}\right),$$
$$P(X>25|\theta) = \exp\left(-\frac{25}{\theta}\right).$$

The likelihood function and the log-likelihood function are given by:

$$L(\theta) = \frac{1}{\theta^3}\exp\left(-\frac{73}{\theta}\right), \quad l(\theta) = -3\ln\theta - \frac{73}{\theta}.$$

Maximizing the likelihood function is difficult. To maximize the log-likelihood function, we can differentiate it with respect to θ and equate it to zero:

$$l'(\theta) = -\frac{3}{\theta} + \frac{73}{\theta^2} = 0.$$

Solving which gives

$$\hat{\theta} = \frac{73}{3}.$$

□

Example 6.2 illustrates how to use the maximum likelihood method to estimate the parameter of an exponential distribution from observed data. In the example, θ is used to denote the true value of the parameter. The estimated value of the parameter is denoted by $\hat{\theta}$.

As we can see from Example 6.2, it is straightforward to handle individual and interval data. Truncated data are common in insurance due to deductibles. Losses below the deductible are not recorded. Handling truncated data in the likelihood function requires more consideration.

In insurance, left truncated data are common. Left truncated data are also called data truncated from below. There are two approaches to handle left truncated data. The first approach is to shift the data by subtracting the truncation point from each observation. The shifted data are used to fit a model. The truncation point is added back to the fitted model later. The second approach is to model the truncated data directly by treating the observations as conditional observations given that the observations are greater than the truncation point. Example 6.3 illustrates the second approach.

Example 6.3. The following five losses are observed from a dataset that is truncated from below at 100:

$$125, \quad 150, \quad 165, \quad 175, \quad 250.$$

An exponential distribution with parameter θ is fitted to the data. Calculate the maximum likelihood estimate of θ.

Solution. Since the dataset is truncated from below at 100, the probabilities of the events are calculated as

$$P(X=125|X>100,\theta) = \frac{P(X=125|\theta)}{P(X>100|\theta)} = \frac{\frac{1}{\theta}\exp\left(-\frac{125}{\theta}\right)}{\exp\left(-\frac{100}{\theta}\right)} = \frac{\exp\left(-\frac{25}{\theta}\right)}{\theta},$$

$$P(X=150|X>100,\theta)=\frac{P(X=150|\theta)}{P(X>100|\theta)}=\frac{\frac{1}{\theta}\exp\left(-\frac{150}{\theta}\right)}{\exp\left(-\frac{100}{\theta}\right)}=\frac{\exp\left(-\frac{50}{\theta}\right)}{\theta},$$

$$P(X=165|X>100,\theta)=\frac{P(X=165|\theta)}{P(X>100|\theta)}=\frac{\frac{1}{\theta}\exp\left(-\frac{165}{\theta}\right)}{\exp\left(-\frac{100}{\theta}\right)}=\frac{\exp\left(-\frac{65}{\theta}\right)}{\theta},$$

$$P(X=175|X>100,\theta)=\frac{P(X=175|\theta)}{P(X>100|\theta)}=\frac{\frac{1}{\theta}\exp\left(-\frac{175}{\theta}\right)}{\exp\left(-\frac{100}{\theta}\right)}=\frac{\exp\left(-\frac{75}{\theta}\right)}{\theta},$$

$$P(X=250|X>100,\theta)=\frac{P(X=250|\theta)}{P(X>100|\theta)}=\frac{\frac{1}{\theta}\exp\left(-\frac{250}{\theta}\right)}{\exp\left(-\frac{100}{\theta}\right)}=\frac{\exp\left(-\frac{150}{\theta}\right)}{\theta}.$$

The likelihood function is

$$L(\theta)=\frac{1}{\theta^5}\exp\left(-\frac{365}{\theta}\right)$$

and the log-likelihood function is

$$l(\theta)=-5\ln\theta-\frac{365}{\theta}.$$

Solving

$$l'(\theta)=-\frac{5}{\theta}+\frac{365}{\theta^2}=0,$$

we get $\hat{\theta}=73$. □

When certain technical conditions (called regularity conditions) are met the maximum likelihood estimator can be shown to have some good asymptotic properties. Theorem 6.1 states that the maximum likelihood estimator of a parameter converges to the normal distribution.

Theorem 6.1 (Asymptotic properties of the maximum likelihood estimator of a single parameter). *Let $\hat{\theta}$ be the maximum likelihood estimator of the parameter θ. Under certain regularity conditions, the distribution of $\sqrt{I(\theta)}(\hat{\theta}-\theta)$ converges to the standard normal distribution, i.e.,*

$$\lim_{n\to\infty}P\left(\sqrt{I(\theta)}(\hat{\theta}-\theta)\le z\right)=\Phi(z),$$

where $I(\theta)$ is the Fisher information given by

$$I(\theta) = -E\left[\frac{\partial^2}{\partial\theta^2} l(\theta)\right] = E\left[\left(\frac{\partial}{\partial\theta} l(\theta)\right)^2\right].$$

Example 6.4. An exponential distribution with parameter θ is fitted to the following five observations:

$$1000, \quad 1400, \quad 5300, \quad 7400, \quad 7600.$$

Calculate the asymptotic variance of the estimate.

Solution. The log-likelihood function corresponding to the maximum likelihood estimation is

$$\begin{aligned} l(\theta) &= \ln\left(\prod_{i=1}^{5} \frac{1}{\theta} \exp\left(-\frac{x_i}{\theta}\right)\right) = -5\ln\theta - \frac{1}{\theta}\sum_{i=1}^{5} x_i \\ &= -5\ln\theta - \frac{22700}{\theta}. \end{aligned}$$

Solving the equation $l'(\theta) = 0$, we get the maximum likelihood estimate $\hat{\theta} = 4540$. Fisher information is calculated as

$$\begin{aligned} I(\theta) &= -E\left[-\frac{\partial^2}{\partial\theta^2} l(\theta)\right] = -E\left[\frac{5}{\theta^2} - \frac{45400}{\theta^3}\right] \\ &= \frac{45400 - 5\theta}{\theta^3}. \end{aligned}$$

Hence the asymptotic variation is

$$I(\hat{\theta})^{-1} = \frac{4540^2}{5} = 4122320.$$

□

When several parameters are estimated by the maximum likelihood method, Theorem 6.2 states the estimator converges to a multivariate normal distribution, which is defined in Definition 6.3.

Theorem 6.2 (Asymptotic properties of the maximum likelihood estimator of several parameters). *Let $\hat{\boldsymbol{\theta}} = (\hat{\theta}_1, \hat{\theta}_2, \ldots, \hat{\theta}_k)^T$ be the maximum likelihood estimator of the vector of parameters $\boldsymbol{\theta}$. Under certain regularity conditions, the distribution of $\hat{\boldsymbol{\theta}} - \boldsymbol{\theta}$ converges to the multivariate normal distribution with mean vector $\mathbf{0}$ and covariance matrix $I^{-1}(\boldsymbol{\theta})$,* *where $I(\boldsymbol{\theta})$ is the Fisher information matrix given by*

$$I(\boldsymbol{\theta}) = -E\begin{bmatrix} \frac{\partial^2 l(\boldsymbol{\theta})}{\partial \theta_1^2} & \frac{\partial^2 l(\boldsymbol{\theta})}{\partial \theta_1 \partial \theta_2} & \cdots & \frac{\partial^2 l(\boldsymbol{\theta})}{\partial \theta_1 \partial \theta_k} \\ \frac{\partial^2 l(\boldsymbol{\theta})}{\partial \theta_2 \partial \theta_1} & \frac{\partial^2 l(\boldsymbol{\theta})}{\partial \theta_2^2} & \cdots & \frac{\partial^2 l(\boldsymbol{\theta})}{\partial \theta_2 \partial \theta_k} \\ \vdots & \vdots & \ddots & \vdots \\ \frac{\partial^2 l(\boldsymbol{\theta})}{\partial \theta_k \partial \theta_1} & \frac{\partial^2 l(\boldsymbol{\theta})}{\partial \theta_k \partial \theta_2} & \cdots & \frac{\partial^2 l(\boldsymbol{\theta})}{\partial \theta_k^2} \end{bmatrix}.$$

Definition 6.3 (Multivariate normal distribution)**.** A random vector $\mathbf{X} = (X_1, X_2, \ldots, X_k)^T$ is said to follow the multivariate normal distribution with mean vector $\boldsymbol{\mu}$ and covariance matrix Σ if it has the following joint pdf

$$f(\mathbf{x}) = \frac{1}{\sqrt{(2\pi)^k \det(\Sigma)}} \exp\left(-\frac{1}{2}(\mathbf{x}-\boldsymbol{\mu})^T \Sigma^{-1}(\mathbf{x}-\boldsymbol{\mu})\right).$$

Example 6.5. The maximum likelihood method is used to fit the Pareto distribution with parameters α and θ to n observations x_1, x_2, ..., x_n. Calculate the asymptotic variances of the estimates $\hat{\alpha}$ and $\hat{\theta}$.

Solution. The log-likelihood function of the Pareto model is

$$\begin{aligned} l(\alpha, \theta) =& \ln\left(\prod_{i=1}^n \frac{\alpha\theta^\alpha}{(x_i+\theta)^{\alpha+1}}\right) = \sum_{i=1}^n (\ln\alpha + \alpha\ln\theta - (\alpha+1)\ln(x_i+\theta)) \\ =& n\ln\alpha + n\alpha\ln\theta - (\alpha+1)\sum_{i=1}^n \ln(x_i+\theta). \end{aligned}$$

The partial derivatives of the log-likelihood function with respect to the parameters are

$$\frac{\partial l(\alpha,\theta)}{\partial\alpha} = \frac{n}{\alpha} + n\ln\theta - \sum_{i=1}^n \ln(x_i+\theta),$$

$$\begin{aligned} \frac{\partial l(\alpha,\theta)}{\partial\theta} =& \frac{n\alpha}{\theta} - (\alpha+1)\sum_{i=1}^n \frac{1}{x_i+\theta}, \\ \frac{\partial^2 l(\alpha,\theta)}{\partial\alpha^2} =& -\frac{n}{\alpha^2}, \end{aligned}$$

$$\frac{\partial^2 l(\alpha,\theta)}{\partial\alpha\partial\theta} =\frac{n}{\theta} - \sum_{i=1}^{n} \frac{1}{x_i+\theta},$$

$$\frac{\partial^2 l(\alpha,\theta)}{\partial\theta^2} = -\frac{n\alpha}{\theta^2} + (\alpha+1)\sum_{i=1}^{n} \frac{1}{(x_i+\theta)^2}.$$

The expectations of the second derivatives are

$$\begin{aligned}
E\left[\frac{\partial^2 l(\alpha,\theta)}{\partial\alpha^2}\right] =& E\left[-\frac{n}{\alpha^2}\right] = -\frac{n}{\alpha^2},\\
E\left[\frac{\partial^2 l(\alpha,\theta)}{\partial\alpha\partial\theta}\right] =& E\left[\frac{n}{\theta} - \sum_{i=1}^{n}\frac{1}{x_i+\theta}\right] = \frac{n}{\theta} - \sum_{i=1}^{n} E\left[\frac{1}{x_i+\theta}\right]\\
=& \frac{n}{\theta} - \sum_{i=1}^{n}\int_0^\infty \frac{1}{x+\theta}\cdot\frac{\alpha\theta^\alpha}{(x+\theta)^{\alpha+1}}\,\mathrm{d}\,x = \frac{n}{\theta} - \frac{n\alpha}{(\alpha+1)\theta}\\
=& \frac{n}{(\alpha+1)\theta},\\
E\left[\frac{\partial^2 l(\alpha,\theta)}{\partial\theta^2}\right] =& E\left[-\frac{n\alpha}{\theta^2} + (\alpha+1)\sum_{i=1}^{n}\frac{1}{(x_i+\theta)^2}\right]\\
=& -\frac{n\alpha}{\theta^2} + (\alpha+1)\sum_{i=1}^{n}\int_0^\infty \frac{1}{(x+\theta)^2}\cdot\frac{\alpha\theta^\alpha}{(x+\theta)^{\alpha+1}}\,\mathrm{d}\,x\\
=& -\frac{n\alpha}{\theta^2} + \frac{n(\alpha+1)\alpha}{(\alpha+2)\theta^2} = -\frac{n\alpha}{(\alpha+2)\theta^2}.
\end{aligned}$$

Hence the Fisher information matrix is

$$I(\alpha,\theta) = \begin{pmatrix} \frac{n}{\alpha^2} & -\frac{n}{(\alpha+1)\theta} \\ -\frac{n}{(\alpha+1)\theta} & \frac{n\alpha}{(\alpha+2)\theta^2} \end{pmatrix}.$$

The inverse of the Fisher information matrix is

$$I^{-1}(\alpha,\theta) = \begin{pmatrix} \frac{\alpha^2(\alpha+1)^2}{n} & \frac{\alpha(\alpha+1)(\alpha+2)\theta}{n} \\ \frac{\alpha(\alpha+1)(\alpha+2)\theta}{n} & \frac{(\alpha+1)^2(\alpha+2)\theta^2}{n\alpha} \end{pmatrix}.$$

By Theorem 6.2, the asymptotic variance of $\hat{\alpha}$ is $\frac{\hat{\alpha}^2(\hat{\alpha}+1)^2}{n}$ and the asymptotic variance of $\hat{\theta}$ is $\frac{(\hat{\alpha}+1)^2(\hat{\alpha}+2)\hat{\theta}^2}{n\hat{\alpha}}$. □

***

Exercise 6.4. An exponential distribution with parameter θ is fitted to the following n observations:

$$x_1, x_2, \ldots, x_n.$$

Show show that maximum likelihood estimate of θ is

$$\hat{\theta} = \bar{x} = \frac{1}{n}\sum_{i=1}^{n} x_i.$$

Exercise 6.5. A gamma distribution with parameter α and θ is used to fit the following data:

$$59, \quad 87, \quad 574, \quad 1204, \quad 1504.$$

Calculate the value of the log-likelihood function at $\alpha = 2$ and $\theta = 1000$.

Exercise 6.6. The following data

$$0.74, \quad 0.81, \quad 0.95$$

are sampled from a distribution with the following pdf:

$$f(x) = (\alpha + 1)x^{\alpha}, \quad 0 < x < 1.$$

Calculate the maximum likelihood estimate of α.

Exercise 6.7. Losses follow the Pareto distribution with parameters α and $\theta = 1$. A random sample of five losses was produced. The values of the three losses are 3, 6 and 14. The values of two losses are greater than 25. Calculate the maximum likelihood estimate of α.

Exercise 6.8. You are given the following random sample of three data points from a Pareto distribution with $\theta = 70$:

$$15 \quad 27 \quad 43$$

Calculate the maximum likelihood estimate for α.

Exercise 6.9. Consider a random sample of 20 losses given below:

Range	Frequency
$[0, 1000]$	7
$[1000, 2000]$	6
$[2000, \infty]$	7

An exponential distribution with parameter θ is fitted to the above data Calculate the maximum likelihood estimate of θ.

Exercise 6.10. A portfolio of policies contains three risk groups. Losses from the low-risk group, the medium-risk group, and the high-risk group follow exponential distributions with parameters θ, 2θ, and 3θ, respectively. A random sample of four losses was produced. Three losses are from the medium-risk group and the values are 1, 2, and 3. One loss is from the high-risk group and the value is 15. Calculate the maximum likelihood estimate of θ.

Exercise 6.11. Ten zeros and ten ones are observed from a count dataset that is truncated from above at 2. A Poisson distribution with parameter λ is used to fit the data. Calculate the maximum likelihood estimate of λ.

Exercise 6.12. The following losses are observed before the deductible is applied:

Loss	Number of losses	Deductible	Policy Limit
200	3	100	∞
300	2	0	1000
> 2000	3	0	2000
400	2	200	∞

A Pareto distribution with parameters α and $\theta = 1000$ is fitted to the data. Calculate the maximum likelihood estimate of α.

Exercise 6.13. The Pareto distribution with parameters $\alpha = 1$ and θ is fitted to the following n observations: x_1, x_2, ..., x_n. The maximum likelihood method is used to estimate θ. Calculate the asymptotic variance of the estimate.

Exercise 6.14. During the past ten years, an insurance policy has experienced the following numbers of claims:

$$10, \quad 2, \quad 4, \quad 0, \quad 6, \quad 2, \quad 4, \quad 5, \quad 4, \quad 2.$$

Suppose that the numbers of claims are independent and follow a Poisson distribution. The maximum likelihood method is used to estimate the parameter of the Poisson distribution. Calculate the asymptotic variance of the estimate.

***

6.3 Bayesian Estimation

The Bayesian method is another popular method for estimating parameters of probability models. In the Bayesian method, the parameters are treated as random variables. A prior distribution is first assumed for the parameters. Then the data is used to update the prior distribution into the posterior distribution, which is used in the prediction. Key components of the Bayesian method are given in Definitions 6.4, 6.5, 6.6, and 6.7.

Definition 6.4 (Prior distribution)**.** The prior distribution is the probability distribution that is assumed for the parameter before some evidence is taken into account. The prior distribution is usually denoted by $\pi(\theta)$.

Definition 6.5 (Model distribution)**.** The model distribution is the conditional probability distribution that is assumed to generate the data given a particular value of the parameter. It is also called the data distribution or sampling distribution. The model distribution is usually denoted by $f_{X|\Theta}(\mathbf{x}|\theta)$.

Given the prior distribution and the model distribution, we can get the joint distribution of the data and the parameter as follows:

$$f(\mathbf{x}, \theta) = f_{X|\Theta}(\mathbf{x}|\theta)\pi(\theta). \tag{6.1}$$

See Section 1.6 for more information about joint distributions. The conditional distribution $f_{X|\Theta}(\mathbf{x}|\theta)$ is the likelihood that is calculated from all the observed data.

By using the Baye's Theorem (see Exercise 1.9), we can get the posterior distribution for the parameter:

$$\pi_{\Theta|X}(\theta|\mathbf{x}) = \frac{f(\mathbf{x}, \theta)}{f_X(\mathbf{x})} = \frac{f_{X|\Theta}(\mathbf{x}|\theta)\pi(\theta)}{f_X(\mathbf{x})}, \tag{6.2}$$

where $f_X(\mathbf{x})$ is the marginal distribution of the data. We can calculate the marginal distribution as follows:

$$f_X(\mathbf{x}) = \int_{\text{supp}(\theta)} f(\mathbf{x}, \theta) \, \mathrm{d}\, \theta = \int_{\text{supp}(\theta)} f_{X|\Theta}(\mathbf{x}|\theta)\pi(\theta) \, \mathrm{d}\, \theta. \tag{6.3}$$

Combining Equation (6.2) and Equation (6.3) yields

$$\pi_{\Theta|X}(\theta|\mathbf{x}) = \frac{f_{X|\Theta}(\mathbf{x}|\theta)\pi(\theta)}{\int_{\text{supp}(\Theta)} f_{X|\Theta}(x|u)\pi(u) \, \mathrm{d}\, u}, \tag{6.4}$$

Definition 6.6 (Posterior distribution)**.** The posterior distribution is the conditional probability distribution of the parameter given the observed data. The posterior distribution is denoted by $\pi_{\Theta|X}(\theta|\mathbf{x})$.

Definition 6.7 (Predictive distribution)**.** The predictive distribution is the conditional probability distribution of a new observation given the data. It is denoted by $f_{Y|X}(y|\mathbf{x})$.

Definition 6.8 (Bayes estimate)**.** The Bayes estimate of a new observation is the mean of the predictive distribution.

Once we have the posterior distribution of the parameter, we can calculate the predictive distribution as follows:

$$f_{Y|X}(y|\mathbf{x}) = \int_{\text{supp}(\Theta)} f_{Y|\Theta}(y|\theta) \cdot \pi_{\Theta|X}(\theta|\mathbf{x}) \,\mathrm{d}\,\theta. \tag{6.5}$$

The Bayes estimate is the mean of the predictive distribution, which is calculated as follows:

$$\begin{aligned} E[Y|\mathbf{x}] &= \int_{\text{supp}(Y)} f_{Y|X}(y|\mathbf{x}) \,\mathrm{d}\,y \\ &= \int_{\text{supp}(Y)} \int_{\text{supp}(\theta)} f_{Y|\Theta}(y|\theta) \cdot \pi_{\Theta|X}(\theta|\mathbf{x}) \,\mathrm{d}\,\theta \,\mathrm{d}\,y \\ &= \int_{\text{supp}(\Theta)} \pi_{\Theta|X}(\theta|\mathbf{x}) \int_{\text{supp}(Y)} f_{Y|\Theta}(y|\theta) \,\mathrm{d}\,y \,\mathrm{d}\,\theta \\ &= \int_{\text{supp}(\Theta)} E[Y|\Theta] \pi_{\Theta|X}(\theta|\mathbf{x}) \,\mathrm{d}\,y. \end{aligned} \tag{6.6}$$

Example 6.6 illustrates how to calculate the Bayes estimate by using the above equation.

Example 6.6. The annual number of claims from a policyholder follows the binomial distribution with parameters $m = 2$ and q, i.e.,

$$P(N = k) = \binom{2}{k} q^k (1-q)^{2-k}, \quad k = 0, 1, 2.$$

The prior distribution of q is assumed to be

$$f(q) = 4q^3, \quad 0 < q < 1.$$

The policyholder filed one claim in Year 1 and one claim in Year 2. Calculate the Bayes estimate of the number of claims in Year 3.

Solution. Since the number of claims is observed for two years, the dataset contains two data points, i.e., $\mathbf{x} = \{1, 1\}$. Given the parameter q, the model distribution is

$$f_{X|\Theta}(\mathbf{x}|q) = \left(\binom{2}{1} q^1 (1-q)^1 \right)^2 = 4q^2(1-q)^2,$$

where Θ and X are the random variables that generate the parameter and the data, respectively. The joint distribution of the data and the parameter is

$$f(\mathbf{x}, q) = f_{X|\Theta}(\mathbf{x}|q)f(q) = 4q^2(1-q)^2 4q^3 = 16q^5(1-q)^2,$$

from which we can derive the posterior distribution for q:

$$\begin{aligned} f_{\Theta|X}(q|\mathbf{x}) =& \frac{f(\mathbf{x}, q)}{\int_0^1 f(\mathbf{x}, u)\,\mathrm{d}\,u} = \frac{16q^5(1-q)^2}{\int_0^1 16u^5(1-u)^2\,\mathrm{d}\,u} \\ =& \frac{q^5(1-q)^2}{B(6,3)} = 168q^5(1-q)^2, \end{aligned}$$

where $B(6, 3)$ is the beta function (see Theorem B.2).

Let Y be the number of claims in Year 3. Since Y follows the binomial distribution with parameters $m = 2$ and q, we have $E[Y|\Theta] = 2q$. By Equation (6.6), we can calculate the Bayes estimate as follows:

$$\begin{aligned} E[Y|\mathbf{x}] =& \int_0^1 2q \cdot 168q^5(1-q)^2\,\mathrm{d}\,q = \int_0^1 336q^6(1-q)^2\,\mathrm{d}\,q \\ =& 336B(7,3) = \frac{336\Gamma(7)\Gamma(3)}{\Gamma(10)} = \frac{4}{3}. \end{aligned}$$

□

The Bayesian estimation method can also be applied when the parameter is discrete. See Example 6.7 for an illustration.

Example 6.7. The annual number of claims from a policyholder in a portfolio of policies follows the following discrete distribution with parameter C. The following table gives the conditional probabilities $P(N = n|C = c)$.

	n				
c	0	1	2	3	4
1	1/3	1/3	1/3	0	0
2	0	1/6	2/3	1/6	0
3	0	0	1/6	2/3	1/6

The prior distribution of C is

$$P(C=1) = \frac{1}{2}, \quad P(C=2) = \frac{1}{3}, \quad P(C=3) = \frac{1}{6}.$$

A randomly selected policyholder filed one claim in Year 1. Calculate the Bayes estimate of the number of claims in Year 2 for this policyholder.

Solution. We have only one data point from the policyholder. Let $\mathbf{x} = \{1\}$ denote the observed data. The joint distribution of N and C is

$$P(N=n, C=c) = P(N=n|C=c)P(C=c), \quad n = 0, 1, \ldots, 4,\ c = 1, 2, 3.$$

The posterior distribution of C can be calculated as follows:

$$\begin{aligned}
P(C=1|\mathbf{x}) &= \frac{P(\mathbf{x}|C=1)P(C=1)}{P(\mathbf{x})} = \frac{P(\mathbf{x}|C=1)P(C=1)}{\sum_{c=1}^{3} P(\mathbf{x}|C=c)P(C=c)} \\
&= \frac{\frac{1}{3}\cdot\frac{1}{2}}{\frac{1}{3}\cdot\frac{1}{2}+\frac{1}{6}\cdot\frac{1}{3}+0\cdot\frac{1}{6}} = \frac{3}{4},
\end{aligned}$$

$$\begin{aligned}
P(C=2|\mathbf{x}) &= \frac{P(\mathbf{x}|C=2)P(C=2)}{P(\mathbf{x})} = \frac{P(\mathbf{x}|C=2)P(C=2)}{\sum_{c=1}^{3} P(\mathbf{x}|C=c)P(C=c)} \\
&= \frac{\frac{1}{6}\cdot\frac{1}{3}}{\frac{1}{3}\cdot\frac{1}{2}+\frac{1}{6}\cdot\frac{1}{3}+0\cdot\frac{1}{6}} = \frac{1}{4}, \\
P(C=3|\mathbf{x}) &= \frac{P(\mathbf{x}|C=3)P(C=3)}{P(\mathbf{x})} = \frac{P(\mathbf{x}|C=3)P(C=3)}{\sum_{c=1}^{3} P(\mathbf{x}|C=c)P(C=c)} \\
&= \frac{0\cdot\frac{1}{6}}{\frac{1}{3}\cdot\frac{1}{2}+\frac{1}{6}\cdot\frac{1}{3}+0\cdot\frac{1}{6}} = 0.
\end{aligned}$$

The expectation of a new observation given the parameter C is

$$\begin{aligned}
E[Y|C=1] &= 1\cdot\frac{1}{3}+2\cdot\frac{1}{3} = 1, \\
E[Y|C=2] &= 1\cdot\frac{1}{6}+2\cdot\frac{2}{3}+3\cdot\frac{1}{6} = 2, \\
E[Y|C=3] &= 2\cdot\frac{1}{6}+3\cdot\frac{2}{3}+4\cdot\frac{1}{6} = 3.
\end{aligned}$$

By Equation (6.6), we get

$$E[Y|\mathbf{x}] = \sum_{c=1}^{3} E[Y|C=c]P(C=c|\mathbf{x}) = 1\cdot\frac{3}{4}+2\cdot\frac{1}{4}+3\cdot 0 = 1.25.$$

□

**

Exercise 6.15. The losses from a portfolio of policies follow the uniform distribution on $[0,\theta]$. The prior distribution of θ is

$$f(\theta) = \frac{500}{\theta^2}, \quad \theta > 500.$$

The following two losses are observed: 400, 600. Calculate the Bayes estimate of the losses from the portfolio.

Exercise 6.16. The losses from a portfolio of policies follow the uniform distribution on $[0, \theta]$. An actuary used the Bayesian method to analyze the loss data and obtained the following posterior distribution for the parameter θ:

$$f_{\Theta|X}(\theta|\mathbf{x}) = \frac{3 \times 600^3}{\theta^4}, \qquad \theta > 600.$$

Calculate the Bayes estimate for the losses.

Exercise 6.17. There are six coins: the first four coins are fair and have a probability of heads 0.5, the fifth coin is biased with a probability of heads 0.25, and the sixth coin is also biased with a probability of heads 0.75. A coin is selected randomly and then flipped repeatedly. The outcome of a flip is recorded as 1 if a head appears or 0 if a tail appears. You observe the following data:

$$1, \quad 1, \quad 0, \quad 1.$$

Calculate the Bayes estimate of the outcome of the next flip.

Exercise 6.18. The annual number of claims from policyholders in a portfolio of policies follows the Poisson distribution with parameter λ. The prior distribution of λ is

$$f(\lambda) = \frac{1}{15}\exp\left(-\frac{\lambda}{6}\right) + \frac{1}{20}\exp\left(-\frac{\lambda}{12}\right), \qquad \lambda > 0.$$

A randomly selected policyholder filed ten claims in Year 1. Calculate the Bayes estimate of the number of claims for this policyholder in Year 2.

Exercise 6.19. The annual number of claims from policyholders in a portfolio of policies follows the Poisson distribution with parameter λ. The prior distribution of λ is

$$f(\lambda) = \lambda^{-2}, \quad \lambda > 1.$$

A random selected policyholder filed five claims in Year 1. Calculate the Bayes estimate of the number of claims for this policyholder in Year 2.

***

7

Model Selection

In this chapter, we introduce several methods to evaluate, compare, and select models. In particular, we introduce graphical methods, hypothesis tests, and information criteria for selecting models. The graphical methods are judgment-based approaches. Hypothesis tests and information criteria are quantitative or score-based approaches.

7.1 Graphical Methods

Graphical methods are the most direct way to see how well a model fits the data. There are several graphs that can be used to compare a model with the data. In this section, we introduce the cdf plot, the P-P plot, and the Q-Q plot, which are the commonly used graphs to compare distributions.

To illustrate these graphs, a synthetic dataset is generated from the exponential distribution with parameter $\theta = 1000$. The synthetic dataset contains 25 data points, which were randomly simulated from the exponential distribution and rounded to ones. Table 7.1 shows the synthetic dataset.

To show how to apply the graphical methods to select models, we fit two models to the synthetic dataset by using the maximum likelihood method. The first model is an exponential distribution with the parameter θ. The maximum likelihood estimate of θ is the mean of the data (see Exercise 6.4):

$$\hat{\theta} = 1030.48.$$

The second model is a gamma distribution with two parameters α and θ. Unlike the maximum likelihood estimate of the exponential distribution, the maximum likelihood estimate of the gamma distribution does not have closed-form formulas. A numerical procedure has to be used to find the optimal values of the parameters that maximize the log-likelihood function. For example, we can use the R code given in Table D.1 to maximize the log-likelihood function. After running the code, we get the following estimates:

$$\hat{\alpha} = 1.3836, \quad \hat{\theta} = 744.4912.$$

Figure 7.1 shows the empirical cdf of the data and the cdfs of the two fitted models. The empirical cdf is defined in Definition 7.1. From the figure, we see

DOI: 10.1201/9781003484899-7

TABLE 7.1
A synthetic dataset generated from the exponential distribution with parameter $\theta = 1000$.

106	140	146	147	294	337	436	540	566	588
642	655	755	762	957	1035	1055	1182	1230	1238
1391	1876	2365	2895	4424					

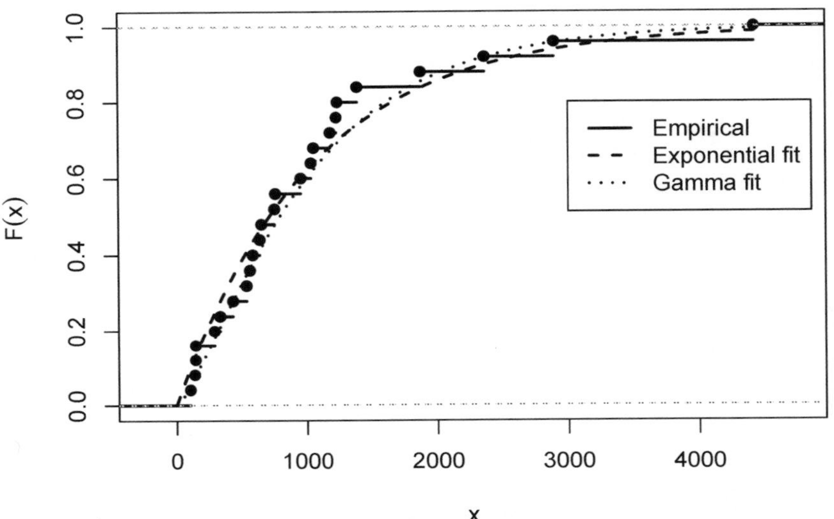

FIGURE 7.1
The empirical cdf and the model cdfs.

that both models fit the data well as the model cdfs match the empirical cdf quite closely. We cannot see the difference between the two models.

Definition 7.1 (Empirical distribution). Let x_1, X_2, ..., x_n be a set of n observations. Then the empirical distribution of this dataset is defined as

$$F_n(x) = \frac{1}{n}\sum_{i=1}^{n} I_{\{x_i \leq x\}},$$

where I is the indicator function.

Figure 7.2 shows the P-P plot of the two fitted models. The P-P plot is a scatter plot between the model cdf and the empirical cdf. That is, a P-P plot shows the coordinates $(F_n(x_i), F(x_i))$ for $i = 1, 2, \ldots, n$. If points are near the line $y = x$ in the P-P plots, then the model fits the data well. From the figure, we see that the points of the two models are around the line $y = x$

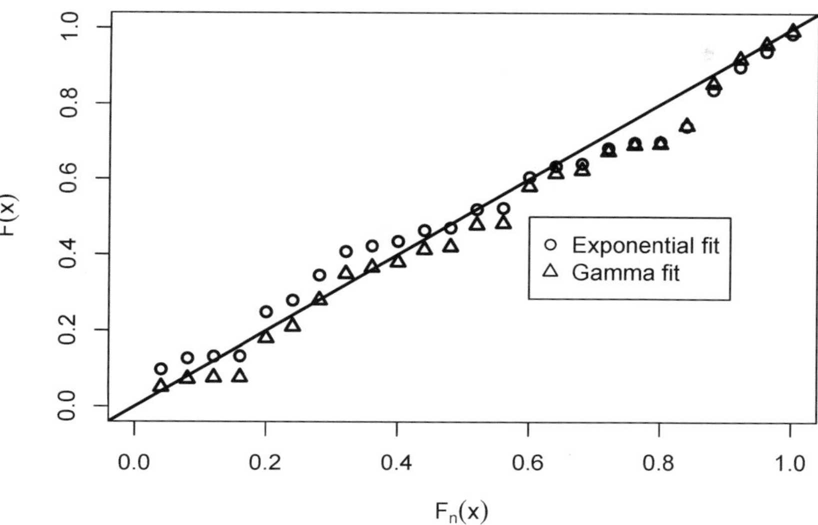

FIGURE 7.2
The P-P plots of the two models. The straight line is $y = x$.

This indicates that both models fit the data well.

Definition 7.2 (Quantile)**.** Let $F(x)$ be the cdf of a distribution. Let $p \in (0, 1)$, then the quantile of the distribution at p is defined as [10]:

$$\pi_p = \inf\{x : F(x) \geq p\}.$$

Quantiles at 1%, 2%, ..., 99% are called percentiles. Quantiles at 25%, 50%, and 75% are called quartiles.

Definition 7.3 (Smoothed empirical quantile)**.** Let $x_1, x_2, \ldots, x_n$ be a set of n observations. Let the order statistics of the data be

$$x_{(1)} \leq x_{(2)} \leq \cdots \leq x_{(n)}.$$

Then the smoothed empirical quantile is defined as

$$\hat{\pi}_p = (1 - h)x_{(j)} + hx_{(j+1)},$$

where $j = \lfloor (n+1)p \rfloor$ and $h = (n+1)p - j$. Here $\lfloor x \rfloor$ denotes the largest integer than is less than or equal to x.

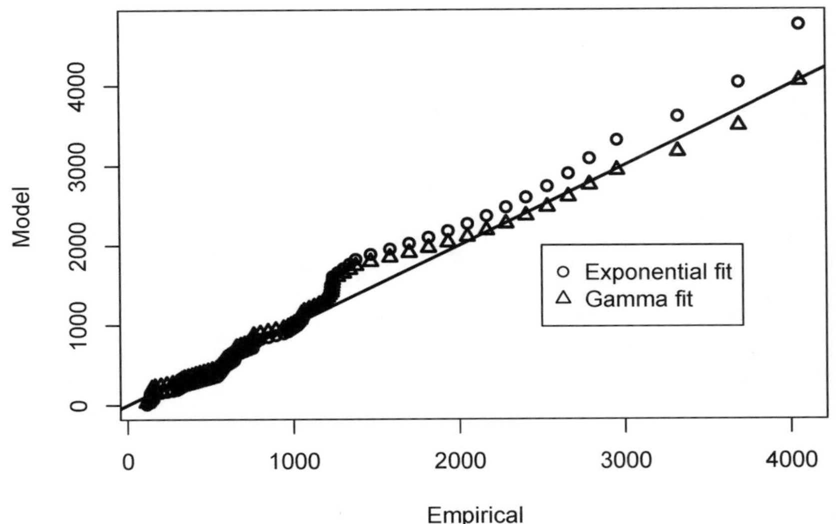

FIGURE 7.3
The Q-Q plots of the two models. The straight line is $y = x$.

The Q-Q plot is a scatter plot between the quantiles of two distributions. The quantile of a distribution is defined in Definition 7.2. The quantile of a sample is defined in Definition 7.3. Figure 7.3 shows the Q-Q plots of the two models. Since the points are near the line $y = x$, the quantiles of the models match well the empirical quantiles of the data. This means that the models fit the data well. Between the two models, the gamma distribution is better because it fits the right tail better than the exponential model.

***---

Exercise 7.1. Consider the following data:

$$1504, \quad 59, \quad 1204, \quad 87, \quad 574.$$

Find the empirical cdf.

Exercise 7.2. Let X follows the exponential distribution with parameter $\theta = 1000$. Calculate the quantile of X at 0.15.

Exercise 7.3. Consider the following data:

$$974, \quad 676, \quad 148, \quad 282, \quad 612, \quad 41, \quad 394, \quad 3202, \quad 514, \quad 183$$

Calculate the 15th smoothed percentile.

Exercise 7.4. You are given the following 20 claims:

$$10, 40, 60, 65, 75, 80, 120, 150, 170, 190, 230, 340, 430,$$
$$440, 980, 600, 675, 950, 1250, 1700$$

An exponential distribution with $\theta = 427.5$ was used to model the data. You are developing a P-P plot for this data. What are the coordinates for $x_7 = 120$.

Exercise 7.5. Consider the following losses from ten randomly selected policies:

$$1, \quad 1, \quad 2, \quad 4, \quad 7, \quad 11, \quad 13, \quad 15, \quad 17, \quad 19.$$

Your colleague fitted a model to the data and the following figure shows the resulting P-P plot:

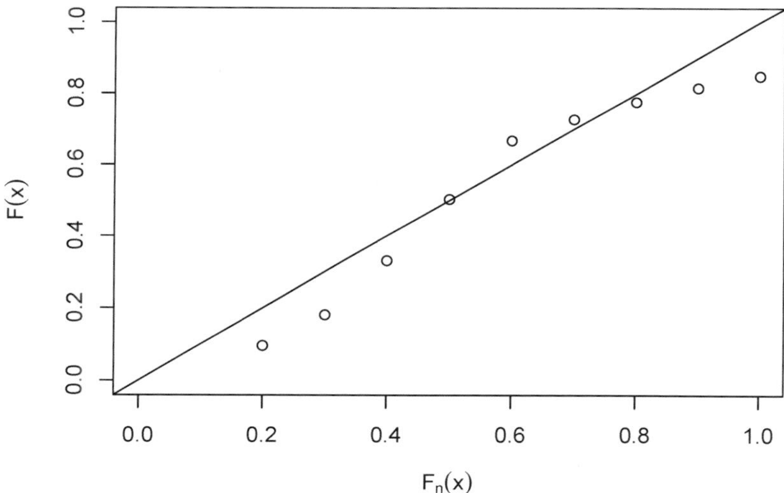

However, your colleague forgot the exact model he fitted. But he remembers that it is one of the following models:

$$F_1(x) = 1 - \exp\left(-\frac{x}{100}\right), \quad F_2(x) = 1 - \exp\left(-\frac{x}{10}\right),$$
$$F_3(x) = 1 - \frac{100}{(x+10)^2}, \quad F_4(x) = \frac{x}{20}.$$

Which is the model your colleague fitted to the data? Justify your answer.

***

TABLE 7.2
Common critical values for the Kolmogorov-Smirnov test.

Significance level	$\alpha = 0.1$	$\alpha = 0.05$	$\alpha = 0.01$
Critical value	$\frac{1.22}{\sqrt{n}}$	$\frac{1.36}{\sqrt{n}}$	$\frac{1.63}{\sqrt{n}}$

7.2 Hypothesis Tests

Hypothesis tests provide a more quantitative way than the graphical methods to compare models. For model comparison purposes, the hypothesis test is usually formulated as:

$$H_0 : \text{The data were generated by the model,}$$
$$H_a : \text{The data were not generated by the model.}$$

Then a test statistic is calculated and a decision is made by comparing the statistic with a critical value corresponding to a significant level. In this section, we introduce several hypothesis tests that are suitable for model comparison.

The first test is the Kolmogorov-Smirnov test, which is defined in Definition 7.4. Since the empirical cdf is a step function, we calculate the test statistic as follows (see Exercise 7.6):

$$D = \max_{1\leq i\leq n} \max(|F_n(x_i-) - F(x_i)|, |F_n(x_i) - F(x_i)|),$$

where x_1, x_2, ..., x_n are the given data points. Table 7.2 gives the common critical values for the Kolmogorov-Smirnov test at different significance levels. The null hypothesis is rejected when the statistic is greater than the critical value.

Definition 7.4 (Kolmogorov-Smirnov test)**.** The Kolmogorov-Smirnov test statistic is defined as

$$D = \max_{d\leq x\leq u} |F_n(x) - F(x)|, \tag{7.1}$$

where d is the left truncation point (or 0 if no truncation), u is the right censoring point (or ∞ if no censoring), $F_n(x)$ is the empirical cdf, and $F(x)$ is the cdf of the model.

Example 7.1. Consider the synthetic dataset given in Table 7.1. Calculate the Kolmogorov-Smirnov test statistics for the exponential model and the gamma model.

TABLE 7.3
Calculation of the Kolmogorov-Smirnov test statistic. Here $F_e(x)$ is the cdf of the exponential model and $M_e(x) = \max(|F_n(x-) - F_e(x)|, |F_n(x) - F_e(x)|)$. The symbols $F_g(x)$ and $M_g(x)$ correspond to the gamma model.

x	$F_n(-x)$	$F_n(x)$	$F_e(x)$	$F_g(x)$	$M_e(x)$	$M_g(x)$
106	0	0.04	0.0978	0.0505	0.0978	0.0505
140	0.04	0.08	0.127	0.0724	0.087	0.0324
146	0.08	0.12	0.1321	0.0763	0.0521	0.0437
147	0.12	0.16	0.1329	0.077	0.0271	0.083
294	0.16	0.2	0.2482	0.1799	0.0882	0.0201
337	0.2	0.24	0.2789	0.2105	0.0789	0.0295
436	0.24	0.28	0.345	0.2797	0.105	0.0397
540	0.28	0.32	0.4079	0.3491	0.1279	0.0691
566	0.32	0.36	0.4226	0.3658	0.1026	0.0458
588	0.36	0.4	0.4348	0.3797	0.0748	0.0203
642	0.4	0.44	0.4637	0.4129	0.0637	0.0271
655	0.44	0.48	0.4704	0.4207	0.0304	0.0593
755	0.48	0.52	0.5194	0.4782	0.0394	0.0418
762	0.52	0.56	0.5226	0.482	0.0374	0.078
957	0.56	0.6	0.6049	0.5803	0.0449	0.0203
1035	0.6	0.64	0.6337	0.6149	0.0337	0.0251
1055	0.64	0.68	0.6408	0.6233	0.0392	0.0567
1182	0.68	0.72	0.6824	0.6733	0.0376	0.0467
1230	0.72	0.76	0.6969	0.6906	0.0631	0.0694
1238	0.76	0.8	0.6992	0.6934	0.1008	0.1066
1391	0.8	0.84	0.7407	0.7427	0.0993	0.0973
1876	0.84	0.88	0.8381	0.8544	0.0419	0.0256
2365	0.88	0.92	0.8992	0.9192	0.0208	0.0392
2895	0.92	0.96	0.9398	0.9578	0.0202	0.0378
4424	0.96	1	0.9863	0.9938	0.0263	0.0338

Solution. From the previous section, we know that the fitted parameter for the exponential model is $\hat{\theta} = 1030.48$ and the fitted parameters for the gamma model are $\hat{\alpha} = 1.3836$ and $\hat{\theta} = 744.4912$. With these parameter values, we can calculate the cdfs of the models and calculate the statistics. The test statistics for the exponential model and the gamma model are given by

$$D_e = 0.1279, \quad D_g = 0.1066,$$

respectively. Table 7.3 shows the detailed calculation. At the 5% significance level, the critical value is $1.36/\sqrt{25} = 0.272$. Since both statistics are less than the critical value, we fail to reject the null hypothesis. This means that both models are plausible models for the data. □

TABLE 7.4
Common critical values for the Anderson-Darling test.

Significance level	$\alpha = 0.1$	$\alpha = 0.05$	$\alpha = 0.01$
Critical value	1.933	2.492	3.857

Definition 7.5 (Anderson-Darling test). The Anderson-Darling test statistic is defined as

$$A^2 = n \int_d^u \frac{(F_n(x) - F(x))^2}{F(x)(1 - F(x))} f(x) \, \mathrm{d}x, \tag{7.2}$$

where (d, u) is the range of the data.

The second test is the Anderson-Darling test, which is defined in Definition 7.5. Since the empirical cdf is a step function, the Anderson-Darling test statistic can also be computed as follows (see Exercise 7.7):

$$A^2 = \sum_{i=0}^{n} S(x_i, x_{i+1}), \tag{7.3}$$

where

$$\begin{aligned} S(x_i, x_{i+1}) =& nF_n^2(x_i) \ln \frac{F(x_{i+1})}{F(x_i)} \\ & - n(F_n(x_i) - 1)^2 \ln \frac{1 - F(x_{i+1})}{1 - F(x_i)} - n(F(x_{i+1}) - F(x_i)). \end{aligned} \tag{7.4}$$

Here $x_0 = 0$. The critical values at different significance levels for the Anderson-Darling test are given in Table 7.4.

Example 7.2. Consider the synthetic dataset given in Table 7.1. Calculate the Anderson-Darling test statistics for the exponential model and the gamma model.

Solution. We can use Equation (7.3) to calculate the test statistics. Table 7.5 shows the details of the calculation. The test statistics for the exponential model and the gamma model are given by

$$D_e = 0.4649, \quad D_g = 0.2789,$$

respectively. At the 5% significance level, the critical value is 2.492. Since both statistics are less than the critical value, we fail to reject the null hypothesis. This means that both models are plausible models for the data. □

TABLE 7.5
Calculation of the Anderson-Darling test statistics. Here $F_e(x)$ is the cdf of the exponential model and $S_e(x_i, x_{i+1})$ is calculated for the exponential model as in Equation (7.4). The symbols $F_g(x)$ and $S_g(x_i, x_{i+1})$ correspond to the gamma model.

i	x_i	$F_n(x_i)$	$F_e(x_i)$	$F_g(x_i)$	$S_e(x_i, x_{i+1})$	$S_g(x_i, x_{i+1})$
0	0	0	0	0	0.1280	0.0330
1	106	0.04	0.0978	0.0505	0.0385	0.0045
2	140	0.08	0.127	0.0724	0.0028	0.0000
3	146	0.12	0.1321	0.0763	0.0000	0.0005
4	147	0.16	0.1329	0.077	0.0342	0.0557
5	294	0.2	0.2482	0.1799	0.0162	0.0005
6	337	0.24	0.2789	0.2105	0.0421	0.0039
7	436	0.28	0.345	0.2797	0.0642	0.0124
8	540	0.32	0.4079	0.3491	0.0138	0.0026
9	566	0.36	0.4226	0.3658	0.0059	0.0003
10	588	0.4	0.4348	0.3797	0.0073	0.0004
11	642	0.44	0.4637	0.4129	0.0005	0.0004
12	655	0.48	0.4704	0.4207	0.0021	0.0071
13	755	0.52	0.5194	0.4782	0.0000	0.0006
14	762	0.56	0.5226	0.482	0.0049	0.0161
15	957	0.6	0.6049	0.5803	0.0014	0.0004
16	1035	0.64	0.6337	0.6149	0.0000	0.0004
17	1055	0.68	0.6408	0.6233	0.0022	0.0066
18	1182	0.72	0.6824	0.6733	0.0016	0.0029
19	1230	0.76	0.6969	0.6906	0.0010	0.0015
20	1238	0.8	0.6992	0.6934	0.0335	0.0418
21	1391	0.84	0.7407	0.7427	0.0457	0.0434
22	1876	0.88	0.8381	0.8544	0.0054	0.0073
23	2365	0.92	0.8992	0.9192	0.0019	0.0094
24	2895	0.96	0.9398	0.9578	0.0094	0.0268
25	4424	1	0.9863	0.9938	0.0024	0.0005
26	∞	1	1	1		

Definition 7.6 (Chi-square goodness-of-fit test)**.** Let $0 = c_0 < c_1 < \ldots < c_k = \infty$ be cutting points. Let

$$E_j = n(F(c_j) - F(c_{j-1})), \quad O_j = n(F_n(c_j) - F_n(c_{j-1})),$$

where n is the number of observations. Then the chi-square goodness-of-fit test statistic is defined as

$$C^2 = \sum_{j=1}^{k} \frac{(E_j - O_j)^2}{E_j}.$$

The test statistic C^2 follows χ^2 distribution with $k - 1 - m$ degrees of freedom, where m is the number of parameters of the model.

The third test is the chi-square goodness-of-fit test, which is defined in Definition 7.6. Unlike the previous two tests, the chi-square goodness-of-fit test requires dividing the range of the distribution into groups. If a model fits the data well, the statistic should be small. The critical value depends on not only the significance level but also the degree of freedom. Table 7.6 shows the critical values for the chi-square goodness-of-fit test.

Example 7.3. Consider the synthetic dataset given in Table 7.1. Calculate the chi-square goodness-of-fit test statistics for the exponential model and the gamma model.

Solution. To calculate the chi-square goodness-of-fit test, we divide the range of the data into six groups. Table 7.7 shows the details of the calculation. The test statistics for the exponential model and the gamma model are given by

$$C_e^2 = 6.011, \quad C_g^2 = 5.5915,$$

respectively. Since we have six groups, the degrees of freedom for the exponential model and the gamma model are 4 and 3, respectively. At the 5% significance level, the critical values are 9.488 and 7.815, respectively. Since both statistics are less than the corresponding critical values, we fail to reject

TABLE 7.6
Common critical values for the chi-square goodness-of-fit test at different degrees of freedom and significance level.

df	$\alpha = 0.1$	$\alpha = 0.05$	$\alpha = 0.01$
1	2.706	3.841	6.635
2	4.605	5.991	9.21
3	6.251	7.815	11.345
4	7.779	9.488	13.277
5	9.236	11.07	15.086
6	10.65	12.592	16.812
7	12.02	14.067	18.475
8	13.36	15.507	20.09
9	14.68	16.919	21.666
10	15.99	18.307	23.209

TABLE 7.7
Calculation of the chi-square goodness-of-fit test.

j	c_j	$F_n(c_j)$	$F(c_j)$	O_j	E_j	$\frac{(E_j - O_j)^2}{E_j}$
			Exponential model			
0	0	0	0			
1	100	0	0.0925	0	2.3125	2.3125
2	200	0.16	0.1764	4	2.0975	1.7256
3	500	0.28	0.3844	3	5.2	0.9308
4	1000	0.6	0.6211	8	5.9175	0.7329
5	2000	0.88	0.8564	7	5.8825	0.2123
6	∞	1	1	3	3.59	0.0970
			Gamma model			
0	0	0	0			
1	100	0	0.0468	0	1.17	1.1700
2	200	0.16	0.1133	4	1.6625	3.2866
3	500	0.28	0.3229	3	5.24	0.9576
4	1000	0.6	0.5997	8	6.92	0.1686
5	2000	0.88	0.8744	7	6.8675	0.0026
6	∞	1	1	3	3.14	0.0062

he null hypothesis. This means that both models are plausible models for the lata. □

The fourth hypothesis test is the likelihood ratio test, which is defined in Definition 7.7. In the likelihood ratio test, the model in the null hypothesis is a special case of the model in the alternative hypothesis. Since the test statistic follows a χ^2 distribution, the critical values given in Table 7.6 can be used to make decisions for the likelihood ratio test.

Definition 7.7 (Likelihood ratio test)**.** The likelihood ratio test statistic is defined as

$$T = 2(l_1(\hat{\boldsymbol{\theta}}_1) - l_0(\hat{\boldsymbol{\theta}}_0)),$$

where $l_0(\cdot)$ and $\hat{\boldsymbol{\theta}}_0$ are the log-likelihood function and the maximum likelihood estimates from the model in the null hypothesis, respectively. Symbols $l_1(\cdot)$ and $\hat{\boldsymbol{\theta}}_1$ correspond to the model in the alternative hypothesis. The statistic T follows a χ^2 distribution with $m_1 - m_0$ degrees of freedom, where m_i is the number of parameters in $\boldsymbol{\theta}_i$ for $i = 0, 1$.

Example 7.4. Consider the synthetic dataset given in Table 7.1. Conduct the following likelihood ratio test:

$$\begin{aligned} H_0 &: \text{The data were generated by a gamma distribution with } \alpha = 1, \\ H_a &: \text{The data were generated by a gamma distribution with } \alpha \neq 1. \end{aligned}$$

Solution. With the help of a computer program, we can get the values of the likelihood functions as given below:

$$l_0(\hat{\boldsymbol{\theta}}_0) = -198.4445, \quad l_1(\hat{\boldsymbol{\theta}}_1) = -197.7079.$$

The test statistic is

$$T = 2(l_1(\hat{\boldsymbol{\theta}}_1) - l_0(\hat{\boldsymbol{\theta}}_0)) = 1.4732.$$

Since the statistic has one degree of freedom, the critical value at 5% significance level is 3.841. Since $T < 3.841$, we fail to reject the null hypothesis. This means that the data were likely generated by the exponential distribution, which is a special case of the gamma distribution with $\alpha = 1$. □

***

Exercise 7.6. Let $x_1, x_2, \ldots, x_n$ be a set of positive data points. Let $F_n(x)$ be the empirical cdf of the data. Let $F(x)$ be the cdf of a distribution with positive support. Show that

$$\max_{0 \le x \le \infty} |F_n(x) - F(x)| = \max_{1 \le i \le n} \max(|F_n(x_i-) - F(x_i)|, |F_n(x_i) - F(x_i)|),$$

where $F_n(x_i-) = \lim_{x \uparrow x_i} F_n(x)$.

Exercise 7.7. Let $x_1, x_2, \ldots, x_n$ be a set of positive data points. Let $F_n(x)$ be the empirical cdf of the data. Let $F(x)$ be the cdf of a distribution with positive support. Show that

$$\begin{aligned} &\int_0^\infty \frac{(F_n(x) - F(x))^2}{F(x)(1 - F(x))} f(x)\, \mathrm{d}x \\ = &-1 + \sum_{i=1}^n F_n^2(x_i) \ln \frac{F(x_{i+1})}{F(x_i)} - \sum_{i=0}^{n-1} (F_n(x_i) - 1)^2 \ln \frac{1 - F(x_{i+1})}{1 - F(x_i)}, \end{aligned}$$

where $x_0 = 0$.

Exercise 7.8. You are given the following 20 claims:

$$\begin{gathered} 10, 40, 60, 65, 75, 80, 120, 150, 170, 190, 230, 340, 430, \\ 440, 980, 600, 675, 950, 1250, 1700 \end{gathered}$$

An exponential distribution with $\theta = 427.5$ was used to model the data Calculate $D(200)$, where $D(x) = |F_n(x) - F(x)|$.

Exercise 7.9. Balog's Bakery has workers' compensation claims during a month of:

$$100, 350, 550, 1000$$

Balog's owner believes that the claims are distributed exponentially with $\theta = 500$. He decides to test his hypothesis at a 10% significance level. Calculate the Kolmogorov-Smirnov test statistic. State the critical value of his test and state his conclusion.

Exercise 7.10. A portfolio of policies has three groups of policyholders. The historical claim probabilities and the current year's claims for the three groups are given below:

Group	Historical Claim Probability	Current Year's Number of Claims
A	0.2744	112
B	0.3512	180
C	0.3744	138

Calculate the chi-square goodness-of-fit test statistic that can be used to test the null hypothesis that the current year's claim probability in each group is the same as the historical probability.

Exercise 7.11. You are given the following data:

Claim	Range Count
0-100	30
100-200	25
200-500	20
500-1000	15
1000+	10

Consider the following hypothesis test:

$$H_0 : \text{The data is from a Pareto distribution,}$$
$$H_a : \text{The data is not from a Pareto distribution.}$$

The parameters were estimated to be $\alpha = 4$ and $\theta = 1200$. Calculate the chi-square test statistic. Calculate the critical value at a 10% significance level. State whether you would reject the Pareto at a 10% significance level.

Exercise 7.12. Based on a random sample, you are testing the following hypothesis:

$$H_0 : \text{The data is from a population distributed binomial with } m = 6, q = 0.3,$$
$$H_1 : \text{The data is from a population distributed binomial.}$$

You are also given the following values of the likelihood function:

$$L(\boldsymbol{\theta}_0) = 0.1, \quad L(\boldsymbol{\theta}_1) = 0.3$$

Calculate the test statistic for the likelihood ratio test. State the critical value at the 10% significance level.

***

7.3 Information Criteria

Like hypothesis tests, information criteria also provide quantitative approaches to select model. Two commonly used information criteria are the Akaike information criterion (AIC) and the Bayesian information criterion (BIC). Definitions 7.8 and 7.9 give the definitions of the two information criteria.

Definition 7.8 (Akaike information criterion)**.** The Akaike information criterion is defined as

$$AIC = 2k - 2l(\hat{\boldsymbol{\theta}}),$$

where k is the number of parameters, $l(\cdot)$ is the log-likelihood function, and $\hat{\boldsymbol{\theta}}$ is the maximum likelihood estimates of the parameters.

Definition 7.9 (Bayesian information criterion)**.** The Bayesian information criterion is defined as

$$BIC = k \ln n - 2l(\hat{\boldsymbol{\theta}}),$$

where n is the number of observations, k is the number of parameters, $l(\cdot)$ is the log-likelihood function, and $\hat{\boldsymbol{\theta}}$ is the maximum likelihood estimates of the parameters. The BIC is also referred to as Schwarz Bayesian criterion (SBC).

The AIC and the BIC are similarly defined. Both criteria penalize the complexity of a model. However, the BIC penalizes the complexity more than does the AIC when n is large. Using the AIC and the BIC to select models is straightforward. Between the two models, we prefer the model with a lower AIC or a lower BIC.

Example 7.5. An exponential model and a gamma model were fitted to the synthetic data given in Table 7.1. The values of the log-likelihood functions are

$$l_e(\hat{\boldsymbol{\theta}}_e) = -198.4445, \quad l_g(\hat{\boldsymbol{\theta}}_g) = -197.7079,$$

where the subscripts e and g indicate the exponential model and the gamma model, respectively. Calculate the AIC and the BIC for the two models.

Solution. The exponential model has one parameter. The gamma model has two parameters. Hence the AIC and the BIC for the two models are given by

$$\begin{aligned} AIC_e &= 2 \times 1 - 2 \times (-198.4445) = 398.889. \\ AIC_g &= 2 \times 2 - 2 \times (-197.7079) = 399.4158, \\ BIC_e &= \ln 25 - 2 \times (-198.4445) = 400.1079, \\ BIC_g &= 2 \ln 25 - 2 \times (-197.7079) = 401.8536. \end{aligned}$$

Both criteria suggest that the exponential model is better than the gamma model. □

***

Exercise 7.13. The following three models were fitted to a dataset:

Model	Number of parameters	Log-likelihood
A	1	-205
B	2	-203
C	3	-200

Calculate the AICs for the three models and state which model is favored by the AIC.

Exercise 7.14. The following three models were fitted to a dataset of 200 observations:

Model	Number of parameters	Log-likelihood
A	1	-205
B	2	-203
C	3	-200

Calculate the BICs for the three models and state which model is favored by the BIC.

Exercise 7.15. Two gamma models are fitted to the following losses:

$$278, \quad 352, \quad 395, \quad 420, \quad 523.$$

The following table gives the estimated parameters of the two models:

	$\hat{\alpha}$	$\hat{\theta}$
Model 1	2	250
Model 2	5	100

Calculate the AICs for the two models and state which model is favored by the AIC.

Exercise 7.16. A model with three parameters is fitted to a dataset. The AIC and the BIC of the fitted model are given below:

$$AIC = 606, \quad BIC = 612.4292.$$

Find the number of observations used to fit the model.

***

7.4 Frequency Model Selection

There is a graphical way to select the discrete models introduced in Chapter 2. Let $\{p_k\}_{k\geq 0}$ be the probability function of a distribution in the $(a, b, 0)$ class. Then the probabilities can be expressed recursively as follows:

$$k\frac{p_k}{p_{k-1}} = ak + b \tag{7.5}$$

for all $k \geq 1$ such that $p_{k-1} > 0$. Given a dataset, we can estimate the left term as follows:

$$k\frac{\hat{p}_k}{\hat{p}_{k-1}} = k\frac{n_k}{n_{k-1}},$$

where n_k is the number of observations whose value is equal to k. Then we plot the ratios $k\dfrac{n_k}{n_{k-1}}$ against k. If one of the $(a, b, 0)$ distributions is a suitable model, then we should see a line that is straight approximately. In this case, the model can be selected based on the slope of the line (see Table 2.1):

(a) If the slope is negative, then the model is the binomial model.

(b) If the slope is near zero, then the model is the Poisson model.

(c) If the slope is positive, then the model is the negative binomial model.

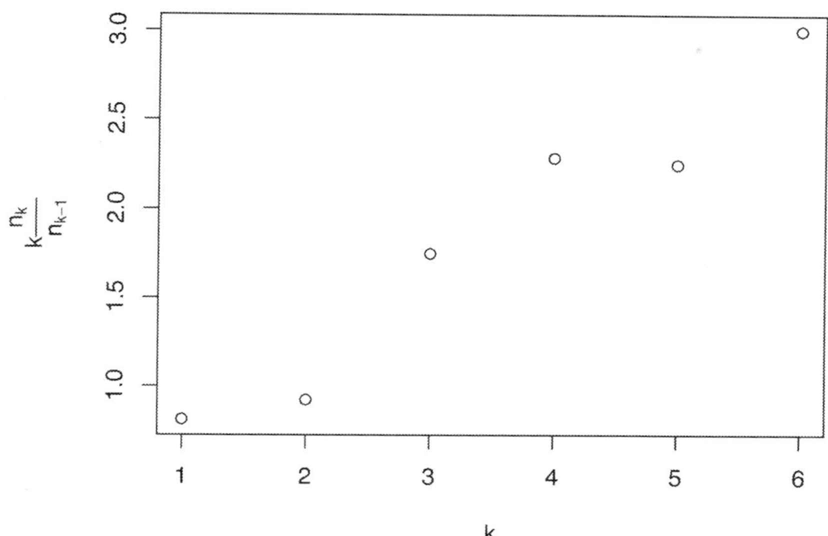

FIGURE 7.4
The scatter plot of the ratio $k\dfrac{n_k}{n_{k-1}}$ against k.

Example 7.6. The following table shows the distribution of accidents for 84 randomly selected policies:

Number of accidents	0	1	2	3	4	5	6
Number of policies	32	26	12	7	4	2	1

Use the graphical approach to select a suitable model for the data.

Solution. We first calculate $k\dfrac{n_k}{n_{k-1}}$ for $k = 1, 2, \ldots, 6$. The following table shows the values:

k	1	2	3	4	5	6
$k\dfrac{n_k}{n_{k-1}}$	0.8125	0.9231	1.75	2.2857	2.25	3

Figure 7.4 shows the scatter plot of the ratio against k. From the figure, we see that the points form approximately a straight line. Hence a model from the $(a, b, 0)$ class is suitable. Since the line has a positive slope, the model is the negative binomial distribution. □

This graphical approach works well when the sample size is large and the data are generated by models in the $(a, b, 0)$ class. A drawback of this graphical approach is that it does not work when $n_k = 0$ for some ks. In such cases, we cannot calculate the ratios.

***

Exercise 7.17. The following table shows the distribution of claims for 100 random selected policies:

Number of claims	0	1	2	3	4	5	6	7
Number of policies	2	9	23	27	26	11	1	1

Use the graphical approach to select a suitable discrete model for the data.

Exercise 7.18. The following table shows the distribution of claims for 300 random selected policies:

Number of claims	0	1	2	3	4	5	6	7	8
Number of policies	68	75	65	33	30	15	8	4	2

Use the graphical approach to select a suitable discrete model for the data.

Exercise 7.19. The following table shows the distribution of claims for 500 random selected policies:

Number of claims	0	1	2	3	4	5	6	7
Number of policies	21	77	125	96	89	58	18	16

Use the graphical approach to select a suitable discrete model for the data.

***

8

Credibility Models

Credibility models play an important role in actuarial science as they are used to set rates in life insurance, health insurance, property and casualty insurance, and pension plans. In this chapter, we introduce several commonly used credibility models.

8.1 Overview

Credibility models were developed in the early 1900s to update the predictions of insurance losses by assigning a numeric weight to the past experience data, which arises from a random process. Consider setting the premium for a risk group, which is a block of insurance policies with similar risk characteristics. Let R be the premium based on the specific risk characteristics of the group. Let M be the manual premium based on a rate specified in the manual. Then the new premium for the next period is set as follows:

$$P_c = wR + (1 - w)M, \tag{8.1}$$

where w is a numeric weight called the credibility factor. When $w = 1$, the past experience is given full credibility. When $0 < w < 1$, the past experience is given partial credibility. When $w = 0$, the past experience is ignored and given no credibility.

The weighted average given in Equation (8.1) can be used to update different measures of claim experience. It can be used to update not just premiums but also the claim frequency, the claim severity, and the aggregate losses. It can also be used to set mortality assumptions, disability assumptions, and other claim rates.

There are three major categories of credibility models to determine the credibility factor w [3]: the limited fluctuation credibility model, the greatest accuracy model, and the Bayesian credibility model. The limited fluctuation credibility model was developed in the early 1900s and used for insurance purposes until the 1960s. If the amount of recent data is sufficient, the limited fluctuation credibility model assigns full credibility (i.e., $w = 1$) to the recent data. If the amount of recent data is insufficient, partial credibility will

DOI: 10.1201/9781003484899-8

be assigned to the recent data. To determine whether the amount of recent data is given full credibility or not, the limited fluctuation credibility model relies on two parameters: the coverage probability and the accuracy parameter. One drawback of the limited fluctuation credibility model is that the two parameters are selected arbitrarily without a rigorous statistical framework.

The greatest accuracy credibility model was developed in the 1960s [4]. Unlike the limited fluctuation credibility model, the greatest accuracy credibility model is formulated within a rigorous statistical framework. As its name indicates, the greatest accuracy credibility model aims to make the estimation errors as small as possible. Examples of the credibility models in this category include the Bühlmann model, the Bülmann-Straub model, and the empirical Bayes model.

The Bayesian credibility model formulates the credibility problem within the Bayesian statistical framework. Such credibility models combine a prior opinion about the unknown quantity and the observed data. The mean of the posterior distribution is used for prediction.

8.2 Limited Fluctuation Credibility

The limited fluctuation credibility model was the oldest credibility model proposed in the early 1900s. It is also known as the classical credibility model and is the most widely used credibility model. The main idea of this model is to assign a higher weight to past experience if the past experience is more stable.

Let X_1, X_2, ..., X_n denote the claim experience (e.g., claim frequency, claim severity, aggregate losses) from the past, where n is the number of observations. Here X_j can be the claim experience from the jth period or the jth policy. The past experience can be summarized by the average:

$$\bar{X} = \frac{1}{n}\sum_{i=1}^{n} X_i. \tag{8.2}$$

Suppose that X_j's are independent with the same mean and the same variance, i.e., $E[X_j] = \mu$ and $\operatorname{Var}(X_j) = \sigma^2$. Then the mean and the variance of $\bar{X}$ are

$$E[(\bar{X})] = \mu, \quad \operatorname{Var}(\bar{X}) = \frac{\sigma^2}{n}.$$

In the fluctuation credibility model, a typical full credibility standard is formulated as follows:

$$P\left(|\bar{X} - \mu| \leq r\mu\right) \geq p, \tag{8.3}$$

where $r > 0$ is close to 0 and $p \in (0,1)$ is close to 1. In other words, if $\bar{X}$ is near the expected value with high probability, then $\bar{X}$ is given full credibility

A common choice for r is 0.05 and a common choice for p is 0.9. Theorem 8.1 shows that full credibility is achieved if the coefficient of variation is not greater than a certain value.

Theorem 8.1. *Suppose that the full credibility standard is given in Equation (8.3). Then full credibility of $\bar{X}$ is achieved if*

$$\frac{\sigma}{\mu} \leq \frac{r\sqrt{n}}{y_p} = \sqrt{\frac{n}{\lambda_0}},$$

where

$$y_p = \inf\left\{y : P\left(\left|\frac{\bar{X}-\mu}{\sigma/\sqrt{n}}\right| \leq y\right) \geq p\right\}, \quad \lambda_0 = \left(\frac{y_p}{r}\right)^2.$$

roof. The criterion given in Equation 8.3 can be written as

$$P\left(\left|\frac{\bar{X}-\mu}{\sigma/\sqrt{n}}\right| \leq \frac{r\mu\sqrt{n}}{\sigma}\right) \geq p.$$

ince y_p is defined to be the minimum value such that $P\left(\left|\frac{\bar{X}-\mu}{\sigma/\sqrt{n}}\right| \leq y\right) \geq p$, ve have

$$\frac{r\mu\sqrt{n}}{\sigma} \geq y_p,$$

hich gives the result. □

When n is large, $\frac{\bar{X}-\mu}{\sigma/\sqrt{n}}$ can be approximated by the standard normal istribution. In such cases, y_p can be obtained from the standard normal istribution as follows:

$$\begin{aligned} p =& \Phi(|Z| \leq y_p) = \Phi(-y_p \leq Z \leq y_p) \\ =& \Phi(y_p) - \Phi(-y_p) = 2\Phi(y_p) - 1, \end{aligned}$$

hich gives $y_p = \Phi^{-1}((p+1)/2)$.

Theorem 8.1 can be applied to different measures of claim experience. xamples 8.1, 8.2, and 8.3 illustrate how to apply this theorem to calculate ne minimum number of observations required to obtain full credibility.

xample 8.1 (Full credibility for claim frequency)**.** The numbers of claims om a portfolio of policies follow a Poisson distribution with parameter λ. he parameter λ is estimated to be 2. The full credibility standard is that ne average number of claims is within 5% of the true value with probability .9. Calculate the minimum number of observations required to obtain full edibility.

Solution. Let $X_1, X_2, \ldots, X_n$ denote the past claim experience. Since X_j's follow the Poisson distribution with parameter λ, we have

$$\mu = E[X_j] = \lambda = 2, \quad \sigma = \sqrt{\operatorname{Var} X_j} = \sqrt{\lambda} = \sqrt{2}.$$

By Theorem 8.1, full credibility is obtained if

$$\sqrt{\frac{n}{\lambda_0}} \geq \frac{\sqrt{2}}{2} = \frac{1}{\sqrt{2}}$$

or

$$n \geq \frac{\lambda_0}{2},$$

where

$$\lambda_0 = \left(\frac{y_p}{r}\right)^2 = \left(\frac{\Phi^{-1}(0.95)}{0.05}\right)^2 = \left(\frac{1.645}{0.05}\right)^2 = 1082.4.$$

Hence

$$n \geq 541.2.$$

The minimum number of observations to obtain full credibility is 542. □

Example 8.2 (Full credibility for claim severity). The claim severity follows the following distribution:

$$f(x) = \frac{1}{10000}, \quad 0 \leq x \leq 10000.$$

The full credibility standard is that the claim severity is within 5% of the true value with probability 0.9. Calculate the number of observations required to obtain full credibility.

Solution. Let $X_1, X_2, \ldots, X_n$ denote the past claim amounts. From the given information, we have

$$\mu = E[X_j] = \int_0^{10000} \frac{x}{10000} \, \mathrm{d}x = 5000,$$

$$\sigma = \sqrt{\operatorname{Var}(X_j)} = \sqrt{E[X_j^2] - E[X_j]^2} = \sqrt{\int_0^{10000} \frac{x^2}{10000} \, \mathrm{d}x - 5000^2}$$

$$= 5000\sqrt{\frac{1}{3}}.$$

By Theorem 8.1, full credibility is obtained if

$$\sqrt{\frac{n}{\lambda_0}} \geq \frac{\sigma}{\mu} = \sqrt{\frac{1}{3}}$$

or

$$n \geq \frac{\lambda_0}{3},$$

where

$$\lambda_0 = \left(\frac{y_p}{r}\right)^2 = \left(\frac{\Phi^{-1}(0.95)}{0.05}\right)^2 = \left(\frac{1.645}{0.05}\right)^2 = 1082.4.$$

Hence

$$n \geq 360.8.$$

The minimum number of observations to obtain full credibility is 361. □

Example 8.3 (Full credibility for aggregate losses)**.** The number of claims from a portfolio of policies follows the Poisson distribution with mean 2. The amount of each claim follows the following distribution:

$$f(x) = \frac{1}{10000}, \quad 0 \leq x \leq 10000.$$

The full credibility standard is that the total cost of claims is within 5% of the true value with probability 0.9. Calculate the number of observations required to obtain full credibility.

Solution. Let X_1, X_2, ..., X_n denote the past claim experience, where the jth observation X_j is the claim amount given by

$$X_j = Y_{j1} + Y_{j2} + \cdots + Y_{jN_j}.$$

Here N_j is the number of claims associated with the jth observation. By Theorem 4.2, we have

$$\begin{aligned}
\mu =& E[X_j] = E[Y]E[N] = 5000 \times 2 = 10000, \\
\sigma =& \sqrt{\operatorname{Var}(X_j)} = \sqrt{E[N]\operatorname{Var}(Y) + \operatorname{Var}(N)E[Y]^2} = \sqrt{2\operatorname{Var}(Y) + 2E[Y]^2} \\
=& \sqrt{2E\left[Y^2\right]} = \sqrt{2\int_0^{10000} \frac{y^2}{10000}\,\mathrm{d}y} = 10000\sqrt{\frac{2}{3}}.
\end{aligned}$$

By Theorem 8.1, full credibility is obtained if

$$\sqrt{\frac{n}{\lambda_0}} \geq \frac{\sigma}{\mu} = \frac{10000\sqrt{\frac{2}{3}}}{10000} = \sqrt{\frac{2}{3}}$$

or

$$n \geq \frac{2\lambda_0}{3},$$

where

$$\lambda_0 = \left(\frac{y_p}{r}\right)^2 = \left(\frac{\Phi^{-1}(0.95)}{0.05}\right)^2 = \left(\frac{1.645}{0.05}\right)^2 = 1082.41.$$

Hence

$$n \geq 721.6.$$

The minimum number of observations to obtain full credibility is 722. □

Example 8.3 illustrates how to find the minimum number of observations of aggregate losses to obtain full credibility when the frequency distribution is known. When the frequency distribution is unknown, we can also use the same approach to derive the conditions for full credibility. The results are summarized in Theorem 8.2.

Theorem 8.2 (Full credibility for aggregate losses). *Let $X_1, X_2, \ldots, X_N$ denote the past losses, where N is the claim frequency random variable. Let $S = X_1 + X_2 + \cdots + X_N$ be the aggregate loss. If the full credibility criterion is that the aggregate loss is within $100r\%$ of the true value with probability p. Then the mean and the standard deviation of the frequency need to satisfy the following conditions:*

$$\frac{\mu_X \mu_N}{\sqrt{\mu_N \sigma_X^2 + \sigma_N^2 \mu_X^2}} \geq \frac{y_p}{r},$$

where μ_N (μ_X)and σ_N (σ_X) denote the mean and the standard deviation of N (X). Here y_p is approximated by

$$y_p = \Phi^{-1}\left(\frac{p+1}{2}\right).$$

In particular, if N follows the Poisson distribution with parameter λ, then the full credibility condition simplifies to

$$\lambda \geq \left(\frac{y_p}{r}\right)^2 \left(1 + \frac{\sigma_X^2}{\mu_X^2}\right).$$

Proof. By Theorem 4.2, we have

$$E[S] = \mu_N \mu_X, \quad \operatorname{Var}(S) = \mu_N \sigma_X^2 + \sigma_N^2 \mu_X^2.$$

The full credibility condition is

$$P(-rE[S] \leq S - E[S] \leq rE[S]) \geq p$$

or

$$P\left(\left|\frac{S - E[S]}{\sqrt{\operatorname{Var}(S)}}\right| \leq \frac{rE[S]}{\sqrt{\operatorname{Var}(S)}}\right) \geq p.$$

Suppose that $\dfrac{S - E[S]}{\sqrt{\operatorname{Var}(S)}}$ is approximately a standard normal random variable then the above condition becomes

$$\frac{rE[S]}{\sqrt{\operatorname{Var}(S)}} \geq \Phi^{-1}(p+1),$$

which is

$$\frac{\mu_X \mu_N}{\sqrt{\mu_N \sigma_X^2 + \sigma_N^2 \mu_X^2}} \geq \frac{y_p}{r}.$$

When N follows the Poisson distribution with parameter λ, we have $\mu_N = \sigma_N^2 = \lambda$. The condition simplifies as given in the theorem. □

Example 8.4. The number of claims from a portfolio of policies follows a Poisson distribution. The amount of each claim follows the following distribution:

$$f(x) = \frac{1}{10000}, \quad 0 \leq x \leq 10000.$$

The full credibility standard is that the total cost of claims is within 5% of the true value with probability 0.9. Calculate the expected number of observations required to obtain full credibility.

Solution. We use Theorem 8.2 to calculate the expected number of observations required to obtain full credibility. Let X be the claim severity random variable. Then

$$\mu_X = E[X] = \int_0^{10000} \frac{x}{10000} \, \mathrm{d}x = 5000$$

and

$$\sigma_X^2 = E[X^2] - E[X]^2 = \int_0^{10000} \frac{x^2}{10000} \, \mathrm{d}x - 5000^2 = \frac{5000^2}{3}.$$

Since the claim frequency follows a Poisson distribution, we apply Theorem 8.2 to obtain the expected number of observations as follows:

$$\lambda \geq \left(\frac{y_p}{r}\right)^2 \left(1 + \frac{\sigma_X^2}{\mu_X^2}\right) = \left(\frac{1.645}{0.05}\right)^2 \left(1 + \frac{1}{3}\right) = 1443.$$

□

When full credibility is not appropriate for the past experience, partial credibility can be given. A simple way to determine partial credibility is the square-root rule for partial credibility. Definition 8.1 gives the square-root rule. We can see that the full credibility standard is a special case of the square-root rule.

Definition 8.1 (Square-root rule for partial credibility). The square-root rule for partial credibility of $\bar{X}$ is defined as

$$w = \min\left\{\frac{\mu}{\sigma}\sqrt{\frac{n}{\lambda_0}}, 1\right\},$$

where λ_0 is defined in Theorem 8.1.

Example 8.5. The claim severity of a block of policies follows the following distribution:

$$f(x) = \frac{1}{10000}, \quad 0 \leq x \leq 10000.$$

The full credibility standard is that the claim severity is within 5% of the true value with probability 0.9. Calculate the partial credibility for 200 observations of claim severity from this block of policies.

Solution. We use the square-root rule given in Definition 8.1 to calculate the partial credibility. The mean and the standard deviation of the claim severity are given by:

$$\mu = \int_0^{10000} \frac{x}{10000} \, \mathrm{d}\, x = 5000,$$

$$\sigma = \sqrt{\int_0^{10000} \frac{x^2}{10000} \, \mathrm{d}\, x - 5000^2} = \frac{5000}{\sqrt{3}}.$$

The value of λ_0 can be obtained from the full credibility standard, i.e.,

$$\lambda_0 = \left(\frac{\Phi^{-1}(0.95)}{0.05} \right)^2 = 1082.278.$$

Hence the partial credibility is

$$w = \frac{5000}{5000/\sqrt{3}} \sqrt{\frac{200}{1082.278}} = 0.7446.$$

□

***

Exercise 8.1. The numbers of claims from a portfolio of policies follow the negative binomial distribution with parameters $r = 2$ and $p = 5/6$. The full credibility standard is that the average number of claims is within 10% of the true value with probability 0.9. Calculate the minimum number of observations required to obtain full credibility.

Exercise 8.2. The numbers of claims from a portfolio of policies follow the negative binomial distribution with parameters $r = 2$ and $p = 5/6$. The amount of each claim follows the Pareto distribution with parameters $\alpha = 3$ and $\theta = 1000$. The full credibility standard is that the aggregate losses from a policy is within 5% of the true value with probability 0.9. Calculate the minimum number of observations required to obtain full credibility.

Exercise 8.3. The claim frequency of a portfolio of policies follows the Poisson distribution with parameter λ. The claim severity follows the gamma distribution with parameters α and θ. The full credibility standard is that the aggregate losses from a policy are within 10% of the true value with probability 0.95. Calculate the minimum number of observations required to obtain full credibility. Express the answer as a function of λ, α, and θ.

Exercise 8.4. The number of claims follows a Poisson distribution with parameter λ. The claim severity follows the Pareto distribution with parameters $\alpha = 6$ and $\theta = 0.5$. The claim frequency and the claim severity are independent. The full credibility criterion is that observed pure premium should be within 2% of the expected pure premium 90% of the time. Calculate the expected number of claims needed for full credibility.

***

8.3 Greatest Accuracy Credibility

Two important greatest accuracy credibility models are the Bühlmann model and the Bühlmann-Straub model. These credibility models calculate the credibility factor by decomposing the variance of the data into two components (see Equation (1.3)).

To describe the Bühlmann model, let $X_1, X_2, \ldots, X_n$ be n observations of losses. These n observations are assumed to be random variables that depend on the parameter Θ, which is also a random variable. The Bühlmann model aims to update the prediction of losses in the next period, denoted by X_{n+1}, based on the n observations by minimizing the prediction error.

Let X denote the common distribution of the observations. By Theorem 1.11, the unconditional variance of the loss random variable X is given by

$$\mathrm{Var}(X) = E[\mathrm{Var}(X|\Theta)] + \mathrm{Var}(E[X|\Theta]).$$

The unconditional variance is called the total variance. The conditional variance $\mathrm{Var}(X|\Theta)$ is called the process variance, which measures the variance of the loss random variable in a given risk group. The conditional expectation $E[X|\Theta]$ is called the hypothetical mean, which measures the mean loss in a given risk group. Table 8.1 summarizes the names of different components.

TABLE 8.1
Names of the components of the unconditional variance.

Notation	Name
$\mathrm{Var}(X)$	Total variance
$\mathrm{Var}(X\|\Theta)$	Process variance
$E[\mathrm{Var}(X\|\Theta)]$	Expected value of the process variance (EPV)
$E[X\|\Theta]$	Hypothetical mean
$\mathrm{Var}(E[X\|\Theta])$	Variance of the hypothetical mean (VHM)

Theorem 8.3 (Bühlmann credibility). *Let X_1, X_2, ..., X_n be n observations of losses. Let X_{n+1} be the loss from the next period. The $n+1$ loss random variables depend on the parameter Θ, which is also a random variable. The random variables X_1, X_2, ..., X_{n+1} are conditionally i.i.d. given Θ. Let*

$$\hat{X}_{n+1} = \beta_0 + \beta_1 X_1 + \cdots \beta_n X_n$$

be a linear predictor of X_{n+1}, where β_0, β_1, ..., β_n are coefficients. The coefficients that minimize the following mean squared error

$$L(\boldsymbol{\beta}) = E\left[\left(X_{n+1} - \hat{X}_{n+1}\right)^2\right]$$

are given by

$$\beta_0 = \frac{k}{n+k} E[X], \quad \beta_1 = \beta_2 = \cdots = \beta_n = \frac{1}{n+k},$$

where $\boldsymbol{\beta} = (\beta_0, \beta_1, \ldots, \beta_n)$ and

$$k = \frac{E[\mathrm{Var}(X|\Theta)]}{\mathrm{Var}(E[X|\Theta])}.$$

The optimal estimate

$$\frac{k}{n+k} E[X] + \frac{X_1 + X_2 + \cdots + X_n}{n+k}$$

is called the Bühlmann credibility estimate. The weight

$$\frac{n}{n+k}$$

given to the sample mean is called the Bühlmann credibility factor.

Proof. Since the mean squared error $L(\boldsymbol{\beta})$ is a quadratic function of the linear coefficients, we can calculate the optimal coefficients by taking the derivatives of $L(\boldsymbol{\beta})$ with respect to the coefficients, equating the derivatives to zeros, and solving the equations. To do that, we first rewrite the mean squared error as

$$L(\boldsymbol{\beta}) = E\left[\left(X_{n+1} - \beta_0 - \sum_{i=1}^{n} \beta_i X_i\right)^2\right].$$

Taking the derivative of $L(\boldsymbol{\beta})$ with respect to β_0, we get

$$\begin{aligned}\frac{\partial L(\boldsymbol{\beta})}{\partial \beta_0} =& E\left[2\left(X_{n+1} - \beta_0 - \sum_{i=1}^{n} \beta_i X_i\right)(-1)\right] \\ =& -2E[X_{n+1}] + 2\beta_0 + 2\sum_{i=1}^{n} \beta_i E[X_i] \\ =& -2E[X] + 2\beta_0 + 2E[X]\sum_{i=1}^{n} \beta_i,\end{aligned}$$

where X is the common distribution shared by X_1, X_2, ..., X_{n+1}. Equating the derivative to zeros gives

$$\beta_0 = E[X]\left(1 - \sum_{i=1}^{n} \beta_i\right). \tag{8.4}$$

For $j = 1, 2, \ldots, n$, the derivative of $L(\boldsymbol{\beta})$ with respect to β_j is

$$\begin{aligned}\frac{\partial L(\boldsymbol{\beta})}{\partial \beta_j} =& E\left[2\left(X_{n+1} - \beta_0 - \sum_{i=1}^{n} \beta_i X_i\right)(-X_j)\right] \\ =& -2E[X_j X_{n+1}] + 2\beta_0 E[X_j] + 2\sum_{i=1}^{n} \beta_i E[X_i X_j]. \qquad (8.5)\end{aligned}$$

Note that

$$E\left[X_j^2\right] = \text{Var}(X_j) + E[X_j]^2 = E[\text{Var}(X|\Theta)] + \text{Var}(E[X|\Theta]) + E[X]^2$$

and for $i \neq j$,

$$\begin{aligned}E[X_i X_j] =& E[E[X_i X_j|\Theta]] = E[E[X_i|\Theta] \cdot E[X_j|\Theta]] \\ =& E\left[E[X|\Theta]^2\right] = \text{Var}(E[X|\Theta]) + E[E[X|\Theta]]^2 \\ =& \text{Var}(E[X|\Theta]) + E[X]^2.\end{aligned}$$

Plugging the above equations and Equation (8.4) into Equation (8.5), we get

$$\begin{aligned}\frac{\partial L(\boldsymbol{\beta})}{\partial \beta_j} =& -2\left(\text{Var}(E[X|\Theta]) + E[X]^2\right) + 2E[X]^2\left(1 - \sum_{i=1}^{n} \beta_i\right) \\ &+ 2\beta_j\left(E[\text{Var}(X|\Theta)] + \text{Var}(E[X|\Theta]) + E[X]^2\right) \\ &+ 2\sum_{i=1, i\neq j}^{n} \beta_i\left(\text{Var}(E[X|\Theta]) + E[X]^2\right) \\ =& -2\,\text{Var}(E[X|\Theta]) + 2\beta_j E[\text{Var}(X|\Theta)] + \sum_{i=1}^{n} \beta_i\,\text{Var}(E[X|\Theta])\end{aligned}$$

for $j = 1, 2, \ldots, n$. Equating the derivatives to zeros gives

$$k\beta_j + \sum_{i=1}^n \beta_i = 1, \quad j = 1, 2, \ldots, n,$$

where $k = E[\mathrm{Var}(X|\Theta)]/\mathrm{Var}(E[X|\Theta])$. It is obvious that

$$\beta_1 = \beta_2 = \cdots = \beta_n = \frac{1}{k}\left(1 - \sum_{i=1}^n \beta_i\right) = \frac{1}{n+k}.$$

Plugging the values of β_1, β_2, ..., β_n into Equation (8.4) gives β_0. □

From Theorem 8.3, we see that the optimal predictor of X_{n+1} is

$$\hat{X}_{n+1} = \frac{k}{n+k}E[X] + \frac{1}{n+k}\sum_{i=1}^n X_i = \frac{k}{n+k}E[X] + \frac{n}{n+k}\bar{X},$$

where $\bar{X}$ is the sample mean of X_1, X_2, ..., X_n, i.e.,

$$\bar{X} = \frac{1}{n}\sum_{i=1}^n X_i.$$

In other words, the optimal prediction is a weighted average of the sample mean and the population mean (e.g., $E[X]$).

Example 8.6. Let X_1, X_2, ..., X_n be n observations of the claim frequency. Suppose that these observations follow the geometric distribution with parameter Θ, which follows $\mathrm{Beta}(\alpha, \beta)$ with $\alpha > 2$. Calculate the Bühlmann credibility factor and the Bühlmann credibility estimate.

Solution. The Bühlmann credibility factor is given by

$$\frac{n}{n+k},$$

where k is given in Theorem 8.3, i.e.,

$$k = \frac{E[\mathrm{Var}(X|\Theta)]}{\mathrm{Var}(E[X|\Theta])}.$$

Since X follows the geometric distribution given Θ, we have

$$E[X|\Theta] = \frac{1-\Theta}{\Theta}, \quad \mathrm{Var}(X|\Theta) = \frac{1-\Theta}{\Theta^2}.$$

Since $\Theta \sim \mathrm{Beta}(\alpha, \beta)$, we have

$$\begin{aligned}
E[\mathrm{Var}(X|\Theta)] =& E\left[\frac{1-\Theta}{\Theta^2}\right] = \int_0^1 \frac{1-\theta}{\theta^2} \cdot \frac{\theta^{\alpha-1}(1-\theta)^{\beta-1}}{B(\alpha,\beta)} \,\mathrm{d}\,\theta \\
=& \int_0^1 \frac{\theta^{\alpha-3}(1-\theta)^{\beta}}{B(\alpha,\beta)} \,\mathrm{d}\,\theta = \frac{B(\alpha-2, \beta+1)}{B(\alpha,\beta)} \\
=& \frac{\Gamma(\alpha-2)\Gamma(\beta+1)}{\Gamma(\alpha+\beta-1)} \cdot \frac{\Gamma(\alpha+\beta)}{\Gamma(\alpha)\Gamma(\beta)} = \frac{\beta(\alpha+\beta-1)}{(\alpha-1)(\alpha-2)}.
\end{aligned}$$

Similarly, we have

$$\begin{aligned} E\left[\frac{1-\Theta}{\Theta}\right] &= \int_0^1 \frac{1-\theta}{\theta}\cdot\frac{\theta^{\alpha-1}(1-\theta)^{\beta-1}}{B(\alpha,\beta)}\,\mathrm{d}\,\theta \\ &= \frac{B(\alpha-1,\beta+1)}{B(\alpha,\beta)} = \frac{\Gamma(\alpha-1)\Gamma(\beta+1)}{\Gamma(\alpha+\beta)}\cdot\frac{\Gamma(\alpha+\beta)}{\Gamma(\alpha)\Gamma(\beta)} \\ &= \frac{\beta}{\alpha-1} \end{aligned}$$

and

$$\begin{aligned} E\left[\left(\frac{1-\Theta}{\Theta}\right)^2\right] &= \int_0^1 \left(\frac{1-\theta}{\theta}\right)^2\cdot\frac{\theta^{\alpha-1}(1-\theta)^{\beta-1}}{B(\alpha,\beta)}\,\mathrm{d}\,\theta \\ &= \frac{B(\alpha-2,\beta+2)}{B(\alpha,\beta)} = \frac{\Gamma(\alpha-2)\Gamma(\beta+2)}{\Gamma(\alpha+\beta)}\cdot\frac{\Gamma(\alpha+\beta)}{\Gamma(\alpha)\Gamma(\beta)} \\ &= \frac{(\beta+1)\beta}{(\alpha-1)(\alpha-2)}. \end{aligned}$$

Hence

$$\begin{aligned} \mathrm{Var}(E[X|\Theta]) = \mathrm{Var}\left(\frac{1-\Theta}{\Theta}\right) &= E\left[\left(\frac{1-\Theta}{\Theta}\right)^2\right] - E\left[\frac{1-\Theta}{\Theta}\right]^2 \\ &= \frac{(\beta+1)\beta}{(\alpha-1)(\alpha-2)} - \left(\frac{\beta}{\alpha-1}\right)^2 = \frac{(\alpha+\beta-1)\beta}{(\alpha-1)^2(\alpha-2)}. \end{aligned}$$

From the above results, we can get the ratio k as follows:

$$k = \frac{\beta(\alpha+\beta-1)}{(\alpha-1)(\alpha-2)}\cdot\frac{(\alpha-1)^2(\alpha-2)}{(\alpha+\beta-1)\beta} = \alpha-1.$$

Hence the Bühlmann credibility factor is

$$\frac{n}{n+\alpha-1}.$$

To calculate the Bühlmann credibility estimate, we need to calculate the population mean. This can be done as follows:

$$E[X] = E[E[X|\Theta]] = E\left[\frac{1-\Theta}{\Theta}\right] = \frac{\beta}{\alpha-1}.$$

Hence the Bühlmann credibility estimate is

$$\begin{aligned} \frac{k}{n+k}E[X] + \frac{n}{n+k}\bar{X} &= \frac{\alpha-1}{n+\alpha-1}\cdot\frac{\beta}{\alpha-1} + \frac{n}{n+\alpha-1}\bar{X} \\ &= \frac{\beta+X_1+X_2+\cdots+X_n}{n+\alpha-1}. \end{aligned}$$

□

In the Bühlmann credibility model, the losses X_1, X_2, ..., X_{n+1} are assumed to be identically distributed. This assumption is usually not satisfied in practice as the losses can be from different periods with different exposures. The Bühlmann-Straub credibility model extends the Bühlmann credibility model by relaxing this assumption. Theorem 8.4 summarizes the Bühlmann-Straub credibility model.

Theorem 8.4 (Bühlmann-Straub credibility). *Let X_1, X_2, ..., X_n be the losses per unit exposure from n periods. Let X_{n+1} be the loss per unit exposure from the next period. For $i = 1, 2, \ldots, n$, let m_i be the exposure in the ith period. The $n+1$ loss random variables depend on the parameter Θ, which is also a random variable. The random variables X_1, X_2, ..., X_{n+1} are conditionally independent given Θ. Given $\Theta = \theta$, the conditional mean and the conditional variance of X_i are given by*

$$E[X_i|\Theta = \theta] = \mu(\theta), \quad \mathrm{Var}(X_i|\Theta = \theta) = \frac{1}{m_i}\sigma^2(\theta),$$

where $\mu(\theta)$ and $\sigma(\theta)$ are suitable functions of θ.

Let

$$\hat{X}_{n+1} = \beta_0 + \beta_1 X_1 + \cdots \beta_n X_n$$

be a linear predictor of X_{n+1}, where β_0, β_1, ..., β_n are coefficients. The coefficients that minimize the following mean squared error

$$L(\boldsymbol{\beta}) = E\left[\left(X_{n+1} - \hat{X}_{n+1}\right)^2\right]$$

are given by

$$\beta_0 = \frac{k}{m_1 + m_2 + \cdots + m_n + k} E[\mu(\Theta)],$$

$$\beta_i = \frac{m_i}{m_1 + m_2 + \cdots + m_n + k}, \qquad i = 1, 2, \ldots, n,$$

where $\boldsymbol{\beta} = (\beta_0, \beta_1, \ldots, \beta_n)$ and

$$k = \frac{E[\sigma^2(\Theta)]}{\mathrm{Var}(\mu(\Theta))}.$$

The weight

$$\frac{m_1 + m_2 + \cdots + m_n}{m_1 + m_2 + \cdots + m_n + k}$$

is called the Bühlmann-Straub credibility factor.

Example 8.7. In a month, the number of claims from any policy in a portfolic follows a Poisson distribution with mean Λ. The claim frequencies of different policies are independent. The parameter Λ follows a gamma distribution:

$$f_\Lambda(\lambda) = \frac{(100\lambda)^6 e^{-100\lambda}}{120\lambda}, \quad \lambda > 0.$$

In Months 1 to 4, the numbers of policies in the portfolios are 100, 150, 200, and 300. In Months 1 to 3, the observed numbers of claims are 6, 8, and 11. Estimate the number of claims in Month 4 by using the Bühlmann-Straub credibility model.

Solution. Let m_i and N_i denote the number of policies and the number of claims in Month i, respectively. Then

$$N_i = N_{i1} + N_{i2} + \cdots + N_{im_i},$$

where N_{ij} is the number of claims from the jth policy. The number of claims per unit exposure is

$$X_i = \frac{N_i}{m_i} = \frac{N_{i1} + N_{i2} + \cdots + N_{im_i}}{m_i}.$$

By the assumption, we have

$$E[X_i|\Lambda = \lambda] = E\left[\frac{N_{i1} + N_{i2} + \cdots + N_{im_i}}{m_i}|\Lambda = \lambda\right] = \lambda$$

and

$$\mathrm{Var}(X_i|\Lambda = \lambda) = \mathrm{Var}\left(\frac{N_{i1} + N_{i2} + \cdots + N_{im_i}}{m_i}|\Lambda = \lambda\right) = \frac{\lambda}{m_i}.$$

Hence we have $\mu(\lambda) = \sigma^2(\lambda) = \lambda$. As a result, the k is

$$k = \frac{E[\sigma^2(\Lambda)]}{\mathrm{Var}(\mu(\Lambda))} = \frac{E[\Lambda]}{\mathrm{Var}(\Lambda)} = \frac{6 \times 0.01}{6 \times 0.01^2} = 100.$$

The Bühlmann-Straub credibility estimate of the number of claims in Month 4 is

$$\begin{aligned} & m_4 \cdot \frac{kE[X] + m_1X_1 + m_2X_2 + m_3X_3}{m_1 + m_2 + m_3 + k} \\ &= 300 \cdot \frac{100 \times 0.06 + 6 + 8 + 11}{100 + 150 + 200 + 100} = 16.91. \end{aligned}$$

□

***

Exercise 8.5. Let X_1, X_2, ..., X_n be n observations of the claim severity. Suppose that these observations follow the exponential distribution with rate parameter Λ:

$$f_X(x) = \Lambda \exp(-\Lambda x).$$

where the rate parameter Λ follows the gamma distribution with parameters α and θ:

$$f_\Lambda(\lambda) = \frac{\lambda^{\alpha-1}\exp(-\lambda/\theta)}{\theta^\alpha \Gamma(\alpha)}.$$

Calculate the Bühlmann credibility factor and the Bühlmann credibility estimate.

Exercise 8.6. The claim experience for a portfolio of policies has the following properties:

(a) The number of claims follows a Poisson distribution with mean θ.

(b) The amount of each claim follows an exponential distribution with mean 10θ.

(c) The claim frequency and the claim severity are conditionally independent given θ.

(d) The distribution of θ has the following pdf

$$f(\theta) = 5\theta^{-6}, \quad \theta > 1.$$

Calculate k for aggregate losses in the Bühlmann credibility model.

Exercise 8.7. A portfolio of policies has two risk groups. In the first risk group, the claim amount follows the following distribution:

$$P(X = 250) = 0.5, \quad P(X = 2500) = 0.3, \quad P(X = 60000) = 0.2.$$

In the second risk group, the claim amount follows the following distribution:

$$P(X = 250) = 0.7, \quad P(X = 2500) = 0.2, \quad P(X = 60000) = 0.1.$$

The claims from the first risk group are twice as likely to be observed than those from the second risk group. A claim of 250 is observed. Estimate the second claim amount by using the Bühlmann credibility model.

Exercise 8.8. The number of claims from a policy in a year follows a Poisson distribution with mean Λ, which is a random variable following a gamma distribution with parameters $\alpha = 1$ and $\theta = 1.2$. In Years 1 and 2, the observed numbers of claims are 3, 0. Estimate the number of claims in Year 3 by using the Bühlmann credibility model.

Exercise 8.9. The claim frequency of an individual policy in a year has mean Λ and variance Σ. It is known that Λ is uniformly distributed on $[0.5, 1.5]$ and Σ is exponentially distributed with mean 1.25. The observed number of claims from a randomly selected policy in Year 1 is zero. Use the Bühlmann credibility model to estimate the number of claims for this policy in Year 2.

Exercise 8.10. Let X denote the claim amount from a policy. The distribution of X depends on a parameter Θ, which is also a random variable. The joint distribution of X and Θ is given by:

$$P(X = 0, \Theta = 0) = 0.4, \quad P(X = 1, \Theta = 0) = 0.1, \quad P(X = 2, \Theta = 0) = 0.1,$$

$$P(X = 0, \Theta = 1) = 0.1 \quad P(X = 1, \Theta = 1) = 0.2, \quad P(X = 2, \Theta = 1) = 0.1.$$

For a given Θ, a sample of size ten for X has a sum of 10, i.e.,

$$\sum_{i=1}^{10} x_i = 10.$$

Calculate the Bühlmann credibility premium.

Exercise 8.11. A portfolio of insurance policies has four classes of policies. Each policy can have 0 or 1 claim with the following distribution:

Class	$P(N = 0)$	$P(N = 1)$
A	0.9	0.1
B	0.8	0.2
C	0.5	0.5
D	0.1	0.9

A class is selected at random with probability $\frac{1}{4}$, and four policies are selected at random from the class. The total number of claims is two. If five policies are selected at random from the same class, estimate the total number of claims by using Bühlmann-Straub credibility model.

Exercise 8.12. You are given the following information about a portfolio of group insurance policies:

(a) Losses for each employee of a given policy are independent and have a common mean and variance.

(b) The overall average loss per employee for all policies is 20.

(c) The variance of the hypothetical means is 40.

(d) The expected value of the process variance is 8000.

(e) The numbers of employees of a randomly selected policy in Years 1, 2, and 3 are 800, 600, and 400, respectively. The losses per employee of this policy in Years 1, 2, and 3 are 15, 10, and 5, respectively.

Calculate the Bühlmann-Straub credibility premium per employee for the selected policy.

***

8.4 Bayesian Credibility

Bayesian credibility models use the Bayesian method to make predictions given the past data. The procedure is the same as the Bayesian estimation outlined in Section 6.3.

Theorem 8.5 (Bayesian credibility). *Let X_1, X_2, ..., X_n be n observations of losses. Let X_{n+1} be the loss from the next period. The $n+1$ loss random variables depend on the parameter Θ, which is also a random variable. The random variables X_1, X_2, ..., X_{n+1} are conditionally independent given Θ. For $i = 1, 2, \ldots, n$, let x_i be a realization of X_i. Let $\mathbf{x} = (x_1, x_2, \ldots, x_n)$. Then the estimator $w(\mathbf{x})$ that minimizes the following mean squared error*

$$L(w(\mathbf{x})) = E\left[\left(E[X_{n+1}|\Theta] - w(\mathbf{x})\right)^2 |\mathbf{x}\right]$$

is given by

$$\hat{w}(\mathbf{x}) = E\left[E[X_{n+1}|\Theta]|\mathbf{x}\right].$$

The optimal estimator $\hat{w}(\mathbf{x})$ is called the Bayes estimator. It is also called the Bayesian premium when X is a loss random variable.

Theorem 8.5 states that the Bayesian estimation is optimal under the squared-error loss function. From Equation (6.6), we see that the Bayesian estimation is the same as the conditional expectation of X_{n+1} given $\mathbf{x}$.

Example 8.8. Let x_1, x_2, ..., x_n be n observed numbers of claims. Suppose that these observed data are realized from the geometric distribution with parameter Θ, which follows Beta(α, β) with $\alpha > 2$. Calculate the Bayes estimate.

Solution. We can use Equation (6.6) to calculate the Bayes estimate. To do that, let $f_{\Theta|X}(\theta|\mathbf{x})$ be the posterior distribution. Then

$$f_{\Theta|X}(\theta|\mathbf{x}) = \frac{f_{X|\Theta}(\mathbf{x}|\theta) f_\Theta(\theta)}{\int_0^1 f_{X|\Theta}(\mathbf{x}|u) f_\Theta(u) \,\mathrm{d}\, u}$$

$$= \frac{\prod_{i=1}^{n}(\theta(1-\theta)^{x_i})\dfrac{\theta^{\alpha-1}(1-\theta)^{\beta-1}}{B(\alpha,\beta)}}{\int_0^1 \prod_{i=1}^{n}(u(1-u)^{x_i})\dfrac{u^{\alpha-1}(1-u)^{\beta-1}}{B(\alpha,\beta)}\,\mathrm{d}\,u}$$
$$= \frac{\theta^{n+\alpha-1}(1-\theta)^{\beta+x_1+x_2+\cdots+x_n-1}}{B(n+\alpha,\beta+x_1+x_2+\cdots+x_n)}.$$

By Equation (6.6), we have

$$\begin{aligned}
E\left[E[X_{n+1}|\Theta]|\mathbf{x}\right] &= \int_0^1 E[X_{n+1}|\Theta=\theta] f_{\Theta|X}(\theta|\mathbf{x})\,\mathrm{d}\,\theta \\
&= \int_0^1 \frac{1-\theta}{\theta}\cdot\frac{\theta^{n+\alpha-1}(1-\theta)^{\beta+x_1+x_2+\cdots+x_n-1}}{B(n+\alpha,\beta+x_1+x_2+\cdots+x_n)}\,\mathrm{d}\,\theta \\
&= \int_0^1 \frac{\theta^{n+\alpha-2}(1-\theta)^{\beta+x_1+x_2+\cdots+x_n}}{B(n+\alpha,\beta+x_1+x_2+\cdots+x_n)}\,\mathrm{d}\,\theta \\
&= \frac{B(n+\alpha-1,\beta+x_1+x_2+\cdots+x_n+1)}{B(n+\alpha,\beta+x_1+x_2+\cdots+x_n)} \\
&= \frac{\Gamma(n+\alpha-1)\Gamma(\beta+x_1+\cdots+x_n+1)}{\Gamma(n+\alpha+\beta+x_1+\cdots+x_n)} \\
&\quad\cdot\frac{\Gamma(n+\alpha+\beta+x_1+\cdots+x_n)}{\Gamma(n+\alpha)\Gamma(\beta+x_1+\cdots+x_n)} \\
&= \frac{\beta+x_1+\cdots+x_n}{n+\alpha-1}.
\end{aligned}$$

□

Comparing Example 8.6 and Example 8.8, we see that the Bühlmann credibility estimate is the same as the Bayes estimate in certain situations.

Example 8.9. The following table shows the data from a credibility model:

First observation	Unconditional probability	Bayes estimate of the second observation
1	1/3	1.5
2	1/3	1.5
3	1/3	3.0

Calculate the Bühlmann credibility estimate of the second observation, given that the first observation is 1.

Solution. Let X denote the observation and let Θ denote the hidden parameter. Let $f_{X,\Theta}(x,\theta)$ denote the joint distribution of X and Θ. Let $f_\Theta(\theta)$ be the marginal distribution of Θ. From the given information, X is discrete and its support is $\{1, 2, 3\}$. We have

$$\begin{aligned}\mu(\theta) =& E[X|\Theta = \theta] = \sum_{x=1}^{3} x f_{X|\Theta}(x|\theta) = \sum_{x=1}^{3} x \frac{f_{X,\Theta}(x,\theta)}{f_\Theta(\theta)} \\ =& \frac{f_{X,\Theta}(1,\theta) + 2f_{X,\Theta}(2,\theta) + 3f_{X,\Theta}(3,\theta)}{f_\Theta(\theta)}\end{aligned}$$

and

$$\sigma^2(\theta) = \mathrm{Var}(X|\Theta = \theta) = E[X^2|\Theta = \theta] - \mu(\theta)^2.$$

We can calculate the first and the second moments of $\mu(\Theta)$ as follows:

$$\begin{aligned}E[\mu(\Theta)] =& \int_{\mathrm{supp}(\Theta)} \mu(\theta) f_\Theta(\theta)\,\mathrm{d}\,\theta = \sum_{x=1}^{3} x \int_{\mathrm{supp}(\Theta)} f_{X,\Theta}(x,\theta)\,\mathrm{d}\,\theta \\ =& \sum_{x=1}^{2} x P(X = x) = \frac{1+2+3}{3} = 2,\end{aligned}$$

$$\begin{aligned}E\left[\mu(\Theta)^2\right] =& \int_{\mathrm{supp}(\Theta)} \mu(\theta)^2 f_\Theta(\theta)\,\mathrm{d}\,\theta = \int_{\mathrm{supp}(\Theta)} \mu(\theta) \left(\sum_{x=1}^{3} x f_{X,\Theta}(x,\theta) \right) \mathrm{d}\,\theta \\ =& \sum_{x=1}^{3} x \int_{\mathrm{supp}(\Theta)} \mu(\theta) f_{X,\Theta}(x,\theta)\,\mathrm{d}\,\theta.\end{aligned}$$

To compute the second moment of $\mu(\Theta)$, we need to use the Bayes estimates. From the given Bayes estimates, we have

$$\begin{aligned}E[\mu(\Theta)|x = 1] =& \int_{\mathrm{supp}(\Theta)} \mu(\theta) f_{\Theta|X}(\theta|1)\,\mathrm{d}\,\theta = \int_{\mathrm{supp}(\Theta)} \mu(\theta) \frac{f_{X,\Theta}(1,\theta)}{P(X=1)}\,\mathrm{d}\,\theta \\ =& 3\int_{\mathrm{supp}(\Theta)} \mu(\theta) f_{X,\Theta}(1,\theta)\,\mathrm{d}\,\theta = 1.5,\end{aligned}$$

which gives

$$\int_{\mathrm{supp}(\Theta)} \mu(\theta) f_{X,\Theta}(1,\theta)\,\mathrm{d}\,\theta = 0.5.$$

Similarly, we have

$$\int_{\mathrm{supp}(\Theta)} \mu(\theta) f_{X,\Theta}(2,\theta)\,\mathrm{d}\,\theta = 0.5.$$

$$\int_{\mathrm{supp}(\Theta)} \mu(\theta) f_{X,\Theta}(3,\theta)\,\mathrm{d}\,\theta = 1.$$

From the above equations, we get

$$E\left[\mu(\theta)^2\right] = 0.5 + 2 \times 0.5 + 3 \times 1 = 4.5.$$

Then we can calculate k as follows:

$$k = \frac{E[\sigma^2(\Theta)]}{\mathrm{Var}(\mu(\Theta))} = \frac{E[X^2] - E[\mu(\Theta)^2]}{E[\mu(\Theta)^2] - E[\mu(\Theta)]^2} = \frac{14/3 - 4.5}{4.5 - 2^2} = \frac{1}{3}.$$

Finally, we can calculate the Bühlmann credibility estimate as follows:

$$\frac{k}{1+k}E[X] + \frac{1}{1+k} \cdot 1 = \frac{2}{4} + \frac{3}{4} = 1.25.$$

□

***

Exercise 8.13. Let $x_1, x_2, \ldots, x_n$ be n observed claim severities. Suppose that these observed data are independent realizations of the exponential distribution with rate parameter Λ:

$$f_X(x) = \Lambda \exp(-\Lambda x).$$

where the rate parameter Λ follows the gamma distribution with parameters α and θ:

$$f_\Lambda(\lambda) = \frac{\lambda^{\alpha-1}\exp(-\lambda/\theta)}{\theta^\alpha \Gamma(\alpha)}.$$

Calculate the Bayes estimate of the claim severity in the next period.

Exercise 8.14. In each year, the number of claims N from a policy follows the following binomial distribution:

$$p_{N|\Theta}(k|q) = \binom{2}{k} q^k (1-q)^{2-k}, \quad k = 0, 1, 2,$$

where Θ is the distribution of the parameter q. The prior distribution of Θ is

$$f_\Theta(q) = 4q^3, \quad q \in (0, 1).$$

In Year 1, the policy has one claim. In Year 2, the policy also has one claim. Calculate the Bayes estimate of the number of claims in Year 3.

Exercise 8.15. Losses from a portfolio of insurance policies follow a Pareto distribution with the following pdf:

$$f_{X|\Theta}(x|\theta) = \frac{2\theta^2}{(x+\theta)^3}, \quad x > 0.$$

The parameter Θ follows the following distribution:

$$P(\Theta = 1) = P(\Theta = 3) = 0.5.$$

The losses of a randomly selected policy in Year 1 were 5. Calculate the Bayes estimate of the losses of the policy in Year 2.

***

8.5 Empirical Bayes Methods

In the Bayesian credibility models, we assume that the distributions $f_{X|\Theta}(\mathbf{x}|\theta)$ and $f_\Theta(\theta)$ and their associated parameters are known. In practice, we need to estimate the parameters from the data. Empirical Bayes methods are procedures that are used to estimate the parameters of $f_{X|\Theta}(\mathbf{x}|\theta)$ and $f_\Theta(\theta)$ from the data.

Empirical Bayes methods can be further classified into nonparametric, semiparametric, and fully parametric methods, depending on how the distributions $f_{X|\Theta}(\mathbf{x}|\theta)$ and $f_\Theta(\theta)$ are specified. When both $f_{X|\Theta}(\mathbf{x}|\theta)$ and $f_\Theta(\theta)$ are unspecified, the method is called the nonparametric empirical Bayes method. When $f_{X|\Theta}(\mathbf{x}|\theta)$ is specified as a parametric form (e.g., the gamma distribution) but $f_\Theta(\theta)$ is unspecified, the method is called the semiparametric empirical Bayes method. When both $f_{X|\Theta}(\mathbf{x}|\theta)$ and $f_\Theta(\theta)$ are assumed to be of parametric forms, the method is called the fully parametric empirical Bayes method.

Theorem 8.6 shows the nonparametric Bühlmann credibility model.

Theorem 8.6 (Nonparametric Bühlmann credibility). *In the Bühlmann credibility model, let r be the number of risk groups. For $i = 1, 2, \ldots, r$, let n_i be the number of loss observations from the ith risk group and let Θ_i be the distribution of the parameter for the ith risk group. The distributions Θ_1, Θ_2, …,, Θ_r are i.i.d. and the common distribution is denoted by Θ. Let X_{ij} be the jth observed loss in the ith risk group. For a fixed i, the observations X_{i1}, X_{i2}, …, X_{in_i} are conditionally independent given Θ_i. For $i \neq s$, the observations X_{ij} and X_{sl} are independent for $j = 1, 2, \ldots, n_i$ and $l = 1, 2, \ldots, n_s$. In addition, the hypothetical means and the process variance are assumed to be*

$$E[X_{ij}|\Theta_i = \theta_i] = \mu_X(\theta_i), \quad \operatorname{Var}(X_{ij}|\Theta_i = \theta_i) = \sigma_X^2(\theta_i).$$

Then

$$\tilde{\mu}_{PV} = \frac{\sum_{i=1}^r \sum_{j=1}^{n_i} (X_{ij} - \bar{X}_i)^2}{\sum_{i=1}^r (n_i - 1)}$$

is an unbiased estimator of the expected value of the process variance $\mu_{PV} = E[\operatorname{Var}(X|\Theta)] = E[\sigma_X^2(\Theta)]$, and

$$\tilde{\sigma}^2_{HM} = \frac{\sum_{i=1}^r n_i(\bar{X}_i - \bar{X})^2 - (r-1)\tilde{\mu}_{PV}}{n - \dfrac{1}{n}\sum_{i=1}^r n_i^2}$$

is an unbiased estimator of the variance of the hypothetical mean $\sigma_{HM}^2 = \mathrm{Var}(E[X|\Theta]) = \mathrm{Var}(\mu_X(\Theta))$. *Here* $\bar{X}_i$ *is the mean observed loss of the* i*th risk group, i.e.,*

$$\bar{X}_i = \frac{1}{n_i} \sum_{j=1}^{n_i} X_{ij},$$

and

$$\bar{X} = \frac{1}{n} \sum_{i=1}^{r} \sum_{j=1}^{n_i} X_{ij} = \frac{1}{n} \sum_{i=1}^{r} n_i \bar{X}_i$$

with $n = \sum_{i=1}^{r} n_i$. *The nonparametric Bühlmann credibility factor for the* i*th risk group is calculated as*

$$\frac{n_i}{n_i + k},$$

where $k = \tilde{\mu}_{PV} / \tilde{\sigma}_{HM}^2$.

Proof. By assumption, we have

$$\begin{aligned} E[X_{ij}^2] =& E[E[X_{ij}^2|\Theta_i]] = E[\mathrm{Var}(X_{ij}|\Theta) + E[X_{ij}|\Theta_i]^2] \\ =& E[\sigma_X^2(\Theta_i)] + E[\mu_X^2(\Theta_i)] = E[\sigma_X^2(\Theta)] + E[\mu_X^2(\Theta)] \end{aligned}$$

and for $j \neq l$,

$$\begin{aligned} E[X_{ij} X_{il}] =& E[E[X_{ij} X_{il}|\Theta_i]] = E[E[X_{ij}|\Theta_i] \cdot E[X_{il}|\Theta_i]] \\ =& E[\mu_X^2(\Theta_i)] = E[\mu_X^2(\Theta)]. \end{aligned}$$

Note that

$$\begin{aligned} \sum_{j=1}^{n_i} (X_{ij} - \bar{X}_i)^2 =& \sum_{j=1}^{n_i} X_{ij}^2 - \frac{1}{n_i} \left(\sum_{j=1}^{n_i} X_{ij} \right)^2 \\ =& \frac{n_i - 1}{n_i} \sum_{j=1}^{n_i} X_{ij}^2 - \frac{1}{n_i} \sum_{1 \leq j \neq l \leq n_i} X_{ij} X_{il} \end{aligned}$$

and

$$\begin{aligned} E[\tilde{\mu}_{PV}] =& E\left[\frac{\sum_{i=1}^{r} \sum_{j=1}^{n_i} (X_{ij} - \bar{X}_i)^2}{\sum_{i=1}^{r} (n_i - 1)} \right] \\ =& \frac{1}{\sum_{i=1}^{r} (n_i - 1)} \sum_{i=1}^{r} E\left[\sum_{j=1}^{n_i} (X_{ij} - \bar{X}_i)^2 \right]. \end{aligned}$$

We have

$$\begin{aligned}
E[\tilde{\mu}_{PV}] =& \frac{1}{\sum_{i=1}^{r}(n_i-1)} \sum_{i=1}^{r}\left(\frac{n_i-1}{n_i}\sum_{j=1}^{n_i}E[X_{ij}^2] - \frac{1}{n_i}\sum_{1\le j\ne l\le n_i}E[X_{ij}X_{il}]\right)\\
=& \frac{1}{\sum_{i=1}^{r}(n_i-1)} \sum_{i=1}^{r}\left((n_i-1)E[\sigma_X^2(\Theta)]\right)\\
=& E[\sigma_X^2(\Theta)] = E[\mathrm{Var}(X|\Theta)] = \mu_{PV},
\end{aligned}$$

which shows that $\tilde{\mu}_{PV}$ is an unbiased estimator of μ_{PV}.

We can also prove that $\tilde{\sigma}_{HM}$ is an unbiased estimator of σ_{HM} in a similar way. To do that, let us first calculate $E[X_{ij}X_{sl}]$ for $s \neq i$:

$$\begin{aligned}
E[X_{ij}X_{sl}] =& E[X_{ij}]E[X_{sl}] = E[E[X_{ij}|\Theta_i]]\cdot E[E[X_{sl}|\Theta_s]]\\
=& E[\mu_X(\Theta_i)]E[\mu_X(\Theta_s)] = E[\mu_X(\Theta)]^2.
\end{aligned}$$

Then we calculate $E[\bar{X}_i\bar{X}_s]$ for $i, s = 1, 2, \ldots, r$. When $i = s$, we have

$$\begin{aligned}
E[\bar{X}_i^2] =& \frac{1}{n_i^2}E\left[\left(\sum_{j=1}^{n_i}X_{ij}\right)^2\right] = \frac{1}{n_i^2}E\left[\sum_{j=1}^{n_i}\sum_{l=1}^{n_i}X_{ij}X_{il}\right]\\
=& \frac{1}{n_i^2}\sum_{j=1}^{n_i}\left(E[\sigma_X^2(\Theta)] + E[\mu_X^2(\Theta)]\right) + \frac{1}{n_i^2}\sum_{1\le j\ne l\le n_i}E[\mu_X^2(\Theta)]\\
=& \frac{1}{n_i}E[\sigma_X^2(\Theta)] + E[\mu_X^2(\Theta)].
\end{aligned}$$

When $i \neq s$, we have

$$\begin{aligned}
E[\bar{X}_i\bar{X}_s] =& \frac{1}{n_i n_s}E\left[\sum_{j=1}^{n_i}\sum_{l=1}^{n_s}X_{ij}X_{sl}\right] = \frac{1}{n_i n_s}\sum_{j=1}^{n_i}\sum_{l=1}^{n_s}E\left[X_{ij}X_{sl}\right]\\
=& \frac{1}{n_i n_s}\sum_{j=1}^{n_i}\sum_{l=1}^{n_s}E[\mu_X(\Theta)]^2 = E[\mu_X(\Theta)]^2.
\end{aligned}$$

From the above result, we have

$$\begin{aligned}
E[\bar{X}_i\bar{X}] =& E\left[\bar{X}_i\frac{1}{n}\sum_{s=1}^{r}n_s\bar{X}_s\right] = \frac{1}{n}\sum_{s=1}^{r}n_sE[\bar{X}_i\bar{X}_s]\\
=& \frac{n_i}{n}E[\bar{X}_i^2] + \frac{1}{n}\sum_{1\le s\le r, s\ne i}n_sE[\bar{X}_i\bar{X}_s]\\
=& \frac{1}{n}E[\sigma_X^2(\Theta)] + \frac{n_i}{n}E[\mu_X^2(\Theta)] + \frac{1}{n}\sum_{1\le s\le r, s\ne i}n_sE[\mu_X(\Theta)]^2\\
=& \frac{1}{n}E[\sigma_X^2(\Theta)] + \frac{n_i}{n}E[\mu_X^2(\Theta)] + \frac{n-n_i}{n}E[\mu_X(\Theta)]^2.
\end{aligned}$$

We also have

$$
\begin{aligned}
E[\bar{X}^2] =& E\left[\left(\frac{1}{n}\sum_{i=1}^{r} n_i\bar{X}_i\right)\bar{X}\right] = \frac{1}{n}\sum_{i=1}^{r} n_i E[\bar{X}_i\bar{X}] \\
=& \frac{1}{n}\sum_{i=1}^{r} n_i\left(\frac{1}{n}E[\sigma_X^2(\Theta)] + \frac{n_i}{n}E[\mu_X^2(\Theta)] + \frac{n-n_i}{n}E[\mu_X(\Theta)]^2\right) \\
=& \frac{1}{n}E[\sigma_X^2(\Theta)] + \frac{\sum_{i=1}^{r} n_i^2}{n^2}E[\mu_X^2(\Theta)] + \frac{n^2-\sum_{i=1}^{r} n_i^2}{n^2}E[\mu_X(\Theta)]^2.
\end{aligned}
$$

From the above equations, we get

$$
\begin{aligned}
\sum_{i=1}^{r} n_i E[(\bar{X}_i-\bar{X})^2] =& \sum_{i=1}^{r} n_i E[\bar{X}_i^2] - 2\sum_{i=1}^{r} n_i E[\bar{X}_i\bar{X}] + \sum_{i=1}^{r} n_i E[\bar{X}^2] \\
=& \sum_{i=1}^{r} n_i E[\bar{X}_i^2] - nE[\bar{X}^2] \\
=& \sum_{i=1}^{r}\left(E[\sigma_X^2(\Theta)] + n_i E[\mu_X^2(\Theta)]\right) - E[\sigma_X^2(\Theta)] \\
& - \frac{\sum_{i=1}^{r} n_i^2}{n}E[\mu_X^2(\Theta)] - \frac{n^2-\sum_{i=1}^{r} n_i^2}{n}E[\mu_X(\Theta)]^2 \\
=& (r-1)E[\sigma_X^2(\Theta)] + \frac{n^2-\sum_{i=1}^{r} n_i^2}{n}\operatorname{Var}(\mu_X(\Theta)).
\end{aligned}
$$

Hence

$$
\begin{aligned}
E[\tilde{\sigma}_{HM}^2] =& E\left[\frac{\sum_{i=1}^{r} n_i(\bar{X}_i-\bar{X})^2 - (r-1)\tilde{\mu}_{PV}}{n - \frac{1}{n}\sum_{i=1}^{r} n_i^2}\right] \\
=& \frac{\sum_{i=1}^{r} n_i E[(\bar{X}_i-\bar{X})^2] - (r-1)E[\sigma_X^2(\Theta)]}{n - \frac{1}{n}\sum_{i=1}^{r} n_i^2} \\
=& \operatorname{Var}(\mu_X(\Theta)) = \sigma_{HM}^2.
\end{aligned}
$$

This completes the proof. □

When $n_1 = n_2 = \cdots = n_r = n_*$, the estimates given in Theorem 8.6 can be simplified as follows:

$$
\tilde{\mu}_{PV} = \frac{\sum_{i=1}^{r}\sum_{j=1}^{n_*}(X_{ij}-\bar{X}_i)^2}{r(n_*-1)},
$$

$$
\tilde{\sigma}_{HM}^2 = \frac{n_*\sum_{i=1}^{r}(\bar{X}_i-\bar{X})^2 - (r-1)\tilde{\mu}_{PV}}{(r-1)n_*}.
$$

Example 8.10. Consider a portfolio of two insurance policies A and B. In Years 1 to 4, the total claims of policy A are 730, 800, 650, and 700 and those of policy B are 655, 650, 625, and 750. Calculate the Bühlmann credibility premium for Policyholder B by using the nonparametric empirical Bayes method.

Solution. We treat each policy as a risk group. Let A be the first risk group and B the second risk group. Then we have $n_1 = n_2 = 4$. Let X_{ij} denote the jth claim amount from the ith risk group. The mean claim amounts are

$$\begin{aligned}\bar{X}_1 =& \frac{730 + 800 + 650 + 700}{4} = 720,\\ \bar{X}_2 =& \frac{655 + 650 + 625 + 750}{4} = 670,\\ \bar{X} =& \frac{4\bar{X}_1 + 4\bar{X}_2}{4+4} = 695.\end{aligned}$$

By Theorem 8.6, we have

$$\begin{aligned}\tilde{\mu}_{PV} =& \frac{10^2 + 80^2 + 70^2 + 20^2 + 15^2 + 20^2 + 45^2 + 80^2}{3+3} = 3475.\\ \tilde{\sigma}^2_{HM} =& 25^2 + 25^2 - \frac{\tilde{\mu}_{PV}}{4} = 381.25.\end{aligned}$$

From the above results, we can calculate k as

$$k = \frac{\tilde{\mu}_{PV}}{\tilde{\sigma}^2_{HM}} = \frac{3475}{381.25} = 9.1148.$$

The Bühlmann credibility premium for Policyholder B is

$$\frac{k}{4+k}\bar{X} + \frac{4}{4+k}\bar{X}_2 = \frac{9.1149}{13.1148} \cdot 695 + \frac{4}{13.1148} \cdot 670 = 687.375.$$

□

Theorem 8.7 gives the nonparametric estimation of the Bühlmann-Straub credibility. The proof is similar to that of Theorem 8.6.

Theorem 8.7 (Nonparametric Bühlmann-Straub credibility). *In the Bühlmann-Straub credibility model, let r be the number of risk groups. For $i = 1, 2, \ldots, r$, let n_i be the number of loss observations from the ith risk group and let Θ_i be the distribution of the parameter for the ith risk group. The distributions Θ_1, Θ_2, $\ldots$,, Θ_r are i.i.d. and the common distribution is denoted by Θ. Let X_{ij} be the jth observed loss per unit exposure in the ith risk group and let m_{ij} be the amount of exposure. For a fixed i, the observations X_{i1}, X_{i2}, $\ldots$, X_{in_i} are conditionally independent given Θ_i. For $i \neq s$, the observations X_{ij} and X_{sl} are independent for $j = 1, 2, \ldots, n_i$ and $l = 1, 2, \ldots, n_s$. In addition, the hypothetical means* *and the process variance are assumed to be*

$$E[X_{ij}|\Theta_i = \theta_i] = \mu_X(\theta_i), \quad \mathrm{Var}(X_{ij}|\Theta_i = \theta_i) = \frac{1}{m_{ij}}\sigma_X^2(\theta_i).$$

Then

$$\tilde{\mu}_{PV} = \frac{\sum_{i=1}^{r}\sum_{j=1}^{n_i} m_{ij}(X_{ij} - \bar{X}_i)^2}{\sum_{i=1}^{r}(n_i - 1)}$$

is an unbiased estimator of the expected value of the process variance $\mu_{PV} = E[\mathrm{Var}(X|\Theta)] = E[\sigma_X^2(\Theta)]$, *and*

$$\tilde{\sigma}_{HM}^2 = \frac{\sum_{i=1}^{r} m_i(\bar{X}_i - \bar{X})^2 - (r-1)\tilde{\mu}_{PV}}{m - \frac{1}{m}\sum_{i=1}^{r} m_i^2}$$

is an unbiased estimator of the variance of the hypothetical mean $\sigma_{HM}^2 = \mathrm{Var}(E[X|\Theta]) = \mathrm{Var}(\mu_X(\Theta))$. *Here* $\bar{X}_i$ *is the mean observed loss of the ith risk group, i.e.,*

$$\bar{X}_i = \frac{1}{m_i}\sum_{j=1}^{n_i} m_{ij}X_{ij},$$

and

$$\bar{X} = \frac{1}{m}\sum_{i=1}^{r}\sum_{j=1}^{n_i} m_{ij}X_{ij} = \frac{1}{m}\sum_{i=1}^{r} m_i\bar{X}_i$$

with $m = \sum_{i=1}^{r} m_i$ *and* $m_i = \sum_{j=1}^{n_i} m_{ij}$. *The nonparametric Bühlmann-Straub credibility factor for the ith risk group is calculated as*

$$\frac{m_i}{m_i + k},$$

where $k = \tilde{\mu}_{PV}/\tilde{\sigma}_{HM}^2$.

In the semiparametric methods, the distribution $f_{X|\Theta}(\mathbf{x}|\theta)$ is specified but the distribution of the parameter Θ is not specified. In this case, we need to estimate the quantities such as $E[\Theta]$ and $\mathrm{Var}(\Theta)$ from the sample data by using the properties of the distribution $f_{X|\Theta}(\mathbf{x}|\theta)$. Example 8.11 illustrates how this is done.

Example 8.11. The number of claims from each policy in a portfolio of motorcycle insurance policies follows a conditional Poisson distribution. In Year 1, the following data are observed:

j	0	1	2	3	4
N_j	2000	600	300	80	20

Here N_j is the number of policies that filed j claims. Calculate the credibility factor for Year 2.

Solution. Let N denote the number of claims from a policy. Let Λ be the parameter of the Poisson distribution that N follows. By the properties of the Poisson distribution, we have

$$E[N|\Lambda] = \Lambda, \quad E[N^2|\Lambda] = \Lambda + \Lambda^2,$$

which gives

$$\Lambda^2 = E[N^2|\Lambda] - \Lambda = E[N^2 - N|\Lambda].$$

Hence

$$\begin{aligned}\tilde{\mu}_{PV} =& E[\text{Var}(N|\Lambda)] = E[\Lambda] = E[E[N|\Lambda]] = E[N] = \bar{N} \\ =& \frac{600 + 2 \times 300 + 3 \times 80 + 4 \times 20}{2000 + 600 + 300 + 80 + 20} = 0.5067.\end{aligned}$$

Similarly,

$$\begin{aligned}\tilde{\sigma}^2_{HM} =& \text{Var}(E[N|\Lambda]) = \text{Var}(\Lambda) = E[\Lambda^2] - E[\Lambda]^2 \\ =& E[E[N^2 - N|\Lambda]] - E[E[N|\Lambda]]^2 \\ =& E[N^2 - N] - E[N]^2 = \text{Var}(N) - E[N] = s^2_N - \bar{N} \\ =& \frac{2000(0 - 0.5067)^2 + 600(1 - 0.5067)^2 + 300(2 - 0.5067)^2}{3000 - 1} \\ &+ \frac{80(3 - 0.5067)^2 + 20(4 - 0.5067)^2}{3000 - 1} - 0.5067 \\ =& 0.6902 - 0.5067 = 0.1835.\end{aligned}$$

The credibility factor is

$$\frac{1}{1+k} = \frac{1}{1 + 0.5067/0.1835} = 0.2659.$$

□

For fully parametric methods, the maximum likelihood methods described in Section 6.2 can be used to estimate the parameters.

***

Exercise 8.16. You are given the following loss data for four policies in seven years:

$$\sum_{i=1}^{4}\sum_{j=1}^{7}(X_{ij} - \bar{X}_i)^2 = 33.6, \quad \sum_{i=1}^{4}(\bar{X}_i - \bar{X})^2 = 3.3,$$

where X_{ij} denotes the loss from the ith policy in the jth year. Use the nonparametric empirical Bayes method to calculate the Bühlmann credibility factor for an individual policy.

Exercise 8.17. The following table shows the data of a commercial auto insurance policies:

Year	Losses	Number of automobiles
	Company A	
1	50,000	100
2	50,000	200
	Company B	
2	150,000	500
3	150,000	300
	Company C	
1	150,000	50
3	150,000	150

Calculate the Bühlmann-Straub credibility factor for Company C.

***

9

Risk Measures

Risk measures play a significant role in pricing, reserving, and risk management in insurance. In this chapter, we provide an overview of risk measures and some common risk measures.

9.1 Overview

As stated in Definition 9.1, a risk measure is a function of a loss distribution that is used to quantify the risk exposure associated with the losses. A risk measure can be thought as the amount of assets required to cover the adverse outcomes of a risk.

Definition 9.1 (Risk measure)**.** A risk measure is a mapping from a loss distribution to the real numbers. Let X be a loss random variable. Then a risk measure for X is denoted by $\rho(X)$.

It is not surprising that there are many risk measures. To help differentiate between risk measures, [2] defined some desirable properties, which include subadditivity, monotonicity, positive homogeneity, and translation invariance. A risk measure with all these four properties are called a coherent measure (see Definition 9.2).

Definition 9.2 (Coherent risk measures)**.** A risk measure $\rho(X)$ is said to be a coherent risk measure if it satisfies the following properties:

(a) $\rho(X+Y) \leq \rho(X) + \rho(Y)$. (Subadditivity)

(b) If $X \leq Y$, then $\rho(X) \leq \rho(Y)$. (Monotonicity)

(c) $\rho(cX) = c\rho(X)$ for all $c > 0$. (Positive homogeneity)

(d) $\rho(X+c) = \rho(X) + c$ for all $c \in (-\infty, \infty)$. (Translation invariance)

The four properties given in Definition 9.2 have intuitive interpretations

DOI: 10.1201/9781003484899-9

Subadditivity means that diversification helps reduce the risk. Monotonicity means that if a risk is always smaller than another risk, then risk measures should be ordered similarly. Positive homogeneity means that risk measures are not affected by changing units of the risk. Translation invariance means that adding a constant amount to a risk will add the same amount to the risk measure.

Example 9.1. Let X be a loss random variable. Show that $\rho(X) = E[X]$ is a coherent risk measure.

Solution. Since $E[X]$ is a mapping from the loss random variable to the real numbers, it is a risk measure. Since the expectation satisfies all the properties given in Definition 9.2, it is a coherent risk measure. □

Example 9.2. Let X be a loss random variable. Let

$$\rho(X) = E[X] + 2\sqrt{\text{Var}(X)}.$$

Show that $\rho(X)$ is not a coherent risk measure.

Solution. According to the properties of the expectation and the variance, this risk measure satisfies the properties of subadditivity, positive homogeneity, and translation invariance. However, it does not satisfy the property of monotonicity. This can be seen from the following example. Let X be a uniform random variable on $[0, 1]$. Then we have

$$E[X] = \frac{1}{2}, \quad \text{Var}(X) = \frac{1}{12}.$$

Let $Y = \frac{1}{2}X + \frac{1}{2}$. Then we have $X \leq Y$. The risk measures for X and Y are given by

$$\begin{aligned} \rho(X) =& E[X] + 2\sqrt{\text{Var}(X)} = \frac{1}{2} + \frac{1}{\sqrt{3}}, \\ \rho(Y) =& \frac{3}{4} + \frac{1}{2\sqrt{3}}. \end{aligned}$$

Hence $\rho(X) > \rho(Y)$. This shows that the risk measure is not coherent. □

***

Exercise 9.1. Let X be a loss random variable. Show that

$$\rho(X) = \frac{1}{\alpha} \ln E\left[e^{\alpha X}\right], \quad \alpha > 0$$

is not a coherent risk measure.

Exercise 9.2. Let X be a loss random variable. Show that

$$\rho(X) = \sqrt{\text{Var}(X)}$$

does not have the monotonicity property.

***

9.2 Value at Risk

The Value at Risk (VaR) is a popular risk measure. It is defined in Definition 9.3. Since it is defined to be a quantile of a random variable, it is also called a quantile risk measure. It can be interpreted as the amount of assets required to ensure the solvency of a company with a certain degree of certainty.

Definition 9.3 (Value at Risk). Let X be a loss random variable. Let $0 < p < 1$. The Value at Risk (VaR) of X at the p level is defined as

$$\mathrm{VaR}_p(X) = \inf\{x \geq 0 : F_X(x) \geq p\},$$

where $F_X(x)$ is the cdf of X.

It can be shown that the VaR has the monotonicity, the positive homogeneity, and the translation invariance properties (see Exercises 9.3, 9.4, 9.5). However, the VaR does not have the subadditivity property. Example 9.3 gives an example showing that the subadditivity property is violated by the VaR.

Example 9.3. Let U_1 and U_2 be uniform random variables on $[0, 1]$. The T_1 and T_2 be discrete random variables with following common distribution

$$P(T = 0) = 0.9, \quad P(T = 10) = 0.1.$$

Suppose that U_1, U_2, T_1, and T_2 are independent. Let $X = U_1 + T_1$ and $Y = U_2 + T_2$. Calculate $\mathrm{VaR}_{0.85}(X)$, $\mathrm{VaR}_{0.85}(Y)$, and $\mathrm{VaR}_{0.85}(X + Y)$.

Solution. The cdf of X is

$$\begin{aligned} F_X(x) &= P(X \leq x) = P(U_1 + T_1 \leq x) \\ &= P(U_1 \leq x)P(T_1 = 0) + P(U_1 \leq x - 10)P(T_1 = 10) \\ &= 0.9P(U_1 \leq x) + 0.1P(U_1 \leq x - 10). \end{aligned}$$

The support of X is $[0, 1] \cup [10, 11]$. To get the VaR of X at 0.95, we need to solve

$$0.9P(U_1 \leq x) + 0.1P(U_1 \leq x - 10) = 0.85.$$

For the above equation to hold, x must be in the interval $[0, 1]$. When $x \in [0, 1]$ the above equation becomes

$$0.9x = 0.85,$$

which gives $x = 0.94$. Hence $\mathrm{VaR}_{0.85}(X) = 0.94$. Since X and Y have the same distribution, we have $\mathrm{VaR}_{0.85}(Y) = 0.94$.

The cdf of $X + Y$ is

$$\begin{aligned}
F_{X+Y}(x) =& P(X + Y \leq x) = P(U_1 + U_2 + T_1 + T_2 \leq x) \\
=& P(U_1 + U_2 \leq x)P(T_1 = 0)P(T_2 = 0) \\
&+ P(U_1 + U_2 \leq x - 1)P(T_1 = 1)P(T_2 = 0) \\
&+ P(U_1 + U_2 \leq x - 1)P(T_1 = 0)P(T_2 = 1) \\
&+ P(U_1 + U_2 \leq x - 2)P(T_1 = 1)P(T_2 = 1) \\
=& 0.81P(U_1 + U_2 \leq x) + 0.18P(U_1 + U_2 \leq x - 10) \\
&+ 0.01P(U_1 + U_2 \leq x - 20).
\end{aligned}$$

The support of $X + Y$ is $[0, 2] \cup [10, 12] \cup [20, 22]$. To get $\mathrm{VaR}_{0.85}(X + Y)$, we need to solve

$$\begin{aligned}
&0.81P(U_1 + U_2 \leq x) + 0.18P(U_1 + U_2 \leq x - 10) \\
&+ 0.01P(U_1 + U_2 \leq x - 20) = 0.85.
\end{aligned}$$

The cdf of $U_1 + U_2$ is (see Example 4.1)

$$F_{U_1+U_2}(x) = \begin{cases} \frac{1}{2}x^2, & \text{if } 0 \leq x \leq 1, \\ -\frac{1}{2}x^2 + 2x - 1, & \text{if } 1 < x \leq 2. \end{cases}$$

For the equation to hold, x must be in the interval $[10, 12]$. When $x \in [10, 12]$, the above equation becomes

$$0.81 + 0.18P(U_1 + U_2 \leq x - 10) = 0.85,$$

or

$$P(U_1 + U_2 \leq x - 10) = \frac{2}{9}.$$

Plugging in to cdf of $U_1 + U_2$, we get

$$\frac{1}{2}(x - 10)^2 = \frac{2}{9}.$$

Solving the above equation gives $x = 10.67$. Hence

$$\mathrm{VaR}_{0.85}(X + Y) = 10.67.$$

□

***---

Exercise 9.3. Let X and Y be loss random variables such that $X \leq Y$. Let $0 < p < 1$. Show that

$$\mathrm{VaR}_p(X) \leq \mathrm{VaR}_p(Y).$$

Exercise 9.4. Let X be a loss random variable and $0 < p < 1$. Show that for any $c > 0$,

$$\mathrm{VaR}_p(cX) = c\,\mathrm{VaR}_p(X).$$

Exercise 9.5. Let X be a loss random variable and $0 < p < 1$. Show that for any $c \in (-\infty, \infty)$,

$$\mathrm{VaR}_p(X + c) = \mathrm{VaR}_p(X) + c.$$

Exercise 9.6. Let U be a uniform random variable on $[0, 1]$. Let T be a discrete random variable with the following pf

$$P(T = 0) = 0.99, \quad P(T = 1) = 0.01.$$

Let $X = U + T$. Calculate $\mathrm{VaR}_{0.95}(X)$.

Exercise 9.7. Let U_1 and U_2 be uniform random variables on $[0, 1]$. The T_1 and T_2 be discrete random variables with the following common distribution

$$P(T = 0) = 0.99, \quad P(T = 1) = 0.01.$$

Suppose that U_1, U_2, T_1, and T_2 are independent. Let $X = U_1 + T_1 + U_2 + T_2$. Calculate $\mathrm{VaR}_{0.99}(X)$.

Exercise 9.8. Let X be a loss random variable with the following pdf:

$$f(x) = \frac{\beta}{1000} \exp\left(-\frac{x}{1000}\right) + \frac{1-\beta}{500} \exp\left(-\frac{x}{500}\right), \quad x > 0,$$

where $\beta \in (0, 1)$. Calculate $\mathrm{VaR}_{0.9}(X)$.

---***

9.3 Tail Value at Risk

The VaR introduced before is used widely in the financial industry when the normal distribution is used to model the asset returns. However, the losses are usually positive and their distributions are skewed. The VaR is generally not used to measure the risk associated with insurance losses. Instead, the tail value at risk (TVaR) is commonly used in the insurance industry. The TVaR is defined in Definition 9.4. The TVaR at level p is the average of all losses that are greater than the VaR at level p. Unlike the VaR, the TVaR is a coherent risk measure. The TVaR is also called the conditional tail expectation (CTE) and is an important risk measure in insurance.

Definition 9.4 (Tail value at risk). Let X be a loss random variable. Let $0 < p < 1$. Suppose that $P(X > \text{VaR}_p(X)) > 0$. Then the tail value at risk of X at the p level is defined as

$$\text{TVaR}_p(X) = E[X|X > \text{VaR}_p(X)].$$

To calculate the TVaR for a discrete random variable, we calculate the conditional expectation. For discrete random variables, the TVaR at level p may use less than the $1 - p$ proportion of the worst losses. See Example 9.4 for an illustration.

Example 9.4. Let X be a loss random variable with the following distribution:

$$P(X = 0) = 0.9, \quad P(X = 10) = 0.05, \quad P(X = 100) = 0.03, \quad P(X = 200) = 0.02.$$

Calculate $\text{TVaR}_{0.95}(X)$ and $\text{TVaR}_{0.97}(X)$.

Solution. First, we need to determine the VaR of X at levels 0.95 and 0.97. By Definition 9.3, we have

$$\begin{aligned}\text{VaR}_{0.95}(X) &= \inf\{x : F_X(x) \geq 0.95\} = 10\\ \text{VaR}_{0.97}(X) &= \inf\{x : F_X(x) \geq 0.97\} = 100.\end{aligned}$$

Then, by Theorem 9.1, we have

$$\begin{aligned}\text{TVaR}_{0.95}(X) &= E[X|X > \text{VaR}_{0.95}(X)] = \frac{1}{0.03 + 0.02}(100 \times 0.03 + 200 \times 0.02)\\ &= 140\end{aligned}$$

and

$$\begin{aligned}\text{TVaR}_{0.97}(X) &= E[X|X > \text{VaR}_{0.97}(X)] = \frac{1}{0.02}(200 \times 0.02)\\ &= 200.\end{aligned}$$

□

To calculate the TVaR for continuous random variables, we can use Theorem 9.1, which states that the TVaR of a continuous random variable X at level p can be obtained from the integration of the VaR values of X at all levels above p.

Theorem 9.1. *If X is a continuous random variable, then the TVaR at level p can be calculated as*

$$\mathrm{TVaR}_p(X) = \frac{1}{1-p}\int_p^1 \mathrm{VaR}_u(X)\,\mathrm{d}\,u.$$

Proof. If X is continuous, then we have $\mathrm{VaR}_p(X) = \pi_p$, which is the quantile at level p. By the change of variable with $y = \pi_u = F_X^{-1}(u)$, we have

$$\begin{aligned}\frac{1}{1-p}\int_p^1 \mathrm{VaR}_u(X)\,\mathrm{d}\,u =& \frac{1}{1-p}\int_{\pi_p}^{\infty} f_X(y)y\,\mathrm{d}\,y\\ =& E[X|X > \pi_p].\end{aligned}$$

□

Example 9.5. Let X be a normal random variable with mean μ and variance σ^2, i.e.,

$$f(x) = \frac{1}{\sqrt{2\pi}\sigma}\exp\left(-\frac{(x-\mu)^2}{2\sigma^2}\right), \quad -\infty < x < \infty.$$

Let $\phi(x)$ and $\Phi(x)$ be the pdf and the cdf of the standard normal distribution, i.e.,

$$\phi(x) = \frac{1}{\sqrt{2\pi}}\exp\left(-\frac{x^2}{2}\right), \quad \Phi(x) = \int_{-\infty}^{x}\phi(u)\,\mathrm{d}\,u.$$

Show that

$$\mathrm{VaR}_p(X) = \mu + \sigma\Phi^{-1}(p), \quad \mathrm{TVaR}_p(X) = \mu + \frac{\sigma}{1-p}\phi(\Phi^{-1}(p)).$$

Solution. Since X is continuous, we have

$$\mathrm{VaR}_p(X) = \pi_p = F_X^{-1}(p).$$

Note that

$$F_X(x) = P(X \le x) = P\left(\frac{X-\mu}{\sigma} \le \frac{x-\mu}{\sigma}\right) = \Phi\left(\frac{x-\mu}{\sigma}\right).$$

Hence π_p can be obtained from

$$F_X(\pi_p) = \Phi\left(\frac{\pi_p-\mu}{\sigma}\right) = p,$$

which gives

$$\pi_p = \mu + \sigma\Phi^{-1}(p).$$

By Theorem 9.1, we have

$$\begin{aligned}
\mathrm{TVaR}_p(X) =& \frac{1}{1-p}\int_{\pi_p}^{\infty} \frac{x}{\sqrt{2\pi}\sigma}\exp\left(-\frac{(x-\mu)^2}{2\sigma^2}\right)\mathrm{d}\,x \\
=& \mu + \frac{1}{1-p}\int_{\pi_p}^{\infty} \frac{x-\mu}{\sqrt{2\pi}\sigma}\exp\left(-\frac{(x-\mu)^2}{2\sigma^2}\right)\mathrm{d}\,x \\
=& \mu + \frac{\sigma}{1-p}\int_{\Phi^{-1}(p)}^{\infty} \frac{y}{\sqrt{2\pi}}\exp\left(-\frac{y^2}{2}\right)\mathrm{d}\,y \\
=& \mu + \frac{\sigma}{1-p}\frac{-1}{\sqrt{2\pi}}\exp\left(-\frac{y^2}{2}\right)\Bigg|_{\Phi^{-1}(p)}^{\infty} = \mu + \frac{\sigma}{1-p}\phi(\Phi^{-1}(p)).
\end{aligned}$$

□

***

Exercise 9.9. Let X follow a Pareto distribution with parameters α and θ. Show that

$$\mathrm{VaR}_p(X) = \theta\left[(1-p)^{-1/\alpha} - 1\right]$$

and

$$\mathrm{TVaR}_p(X) = \frac{\alpha}{\alpha-1}\,\mathrm{VaR}_p(X) + \frac{\theta}{\alpha-1}.$$

Exercise 9.10. Let X have the following pdf

$$f(x) = \frac{1}{b-a}, \quad x \in [a, b],$$

where $a < b$ are constants. Determine the following

(a) $E[X^k]$

(b) $\mathrm{Var}(X)$

(c) $e(d)$, where $d \in [a, b]$

(d) $\mathrm{VaR}_p(X)$

(e) $\mathrm{TVaR}_p(X)$

Exercise 9.11. Let X be a loss random variable following the exponential distribution with parameter θ. Let $0 < p < 1$. Calculate $\mathrm{TVaR}_p(X) - \mathrm{VaR}_p(X)$.

Exercise 9.12. Let X and Y be two random variables with the following joint pdf:

$$f(x, y) = 2, \quad 0 < x < 1,\ 0 < y < 1,\ 0 < x + y < 1.$$

Let $S = X + Y$. Calculate $\mathrm{TVaR}_{0.75}(S)$.

Exercise 9.13. A loss random variable X follows the following distribution:

$$P(X = i) = \frac{1001 - i}{500500}, \quad i = 1, 2, \ldots, 1000.$$

You are given

$$\sum_{i=1}^{n} i = \frac{n(n+1)}{2}, \quad n = 1, 2, \ldots.$$

$$\sum_{i=1}^{n} i^2 = \frac{n(n+1)(2n+1)}{6}, \quad n = 1, 2, \ldots.$$

Calculate $\mathrm{TVaR}_{0.75}(X)$.

Exercise 9.14. Let X be a random variable with the following pdf:

$$f(x) = \frac{1000 - x}{500000}, \quad 0 < x \leq 1000.$$

Calculate $\mathrm{TVaR}_{0.75}(X)$.

---***

Part I

Appendix

A

Useful Results from Calculus

Calculus is a branch of mathematics that deals with variable quantities. In fact, calculus provides a general framework for analyzing variable quantities. In this section, we present some commonly used results from calculus. In particular, we present results related to derivatives, limits, and integration as well as the definition of some important functions. Readers are referred to [9] for further details.

Theorem A.1 (Basic Properties of Integration). *Let $f(x)$, $g(x)$, and $h(x)$ be functions. The definite integral has the following properties:*

(a) $\int_a^b f(x)\,\mathrm{d}\,x + \int_b^c f(x)\,\mathrm{d}\,x = \int_a^c f(x)\,\mathrm{d}\,x$.

(b) If $g(x) \le f(x) \le h(x)$ *for all* $x \in [a,b]$, *then* $\int_a^b g(x)\,\mathrm{d}\,x \le \int_a^b f(x)\,\mathrm{d}\,x \le \int_a^b h(x)\,\mathrm{d}\,x$.

(c) $\int_a^b [f(x) + g(x)]\,\mathrm{d}\,x = \int_a^b f(x)\,\mathrm{d}\,x + \int_a^b g(x)\,\mathrm{d}\,x$.

(d) If c is a constant, then $\int_a^b c \cdot f(x)\,\mathrm{d}\,x = c \cdot \int_a^b f(x)\,\mathrm{d}\,x$.

Theorem A.2 (Arithmetic Properties of Derivatives). *Let $f(x)$ and $g(x)$ be two differentiable functions on (a,b). Let c be a real number. Then for $x \in (a,b)$,*

(a) $(f+g)'(x) = f'(x) + g'(x)$.

(b) $(cf)'(x) = cf'(x)$.

(c) $(fg)'(x) = f'(x)g(x) + f(x)g'(x)$.

(d) $\left(\frac{f}{g}\right)'(x) = \frac{f'(x)g(x) - f(x)g'(x)}{g(x)^2}$ *if* $g(x) \neq 0$.

DOI: 10.1201/9781003484899-A

Theorem A.3 (Chain Rule). *Let f and g be differentiable on (a, b) and (c, d), respectively. If $f(x) \in (c, d)$ for all $x \in (a, b)$, then, $g \circ f$ is differentiable on (a, b) with*

$$(g \circ f)'(x) = g'(f(x))f'(x), \quad x \in (a, b).$$

Here $(g \circ f)(x) = g(f(x))$.

Theorem A.4 (Fundamental Theorem of Analysis). *Let $f(x)$ be a continuous function defined on $[a, b]$. Let $F(x) = \int_a^x f(u)\,\mathrm{d}\,u$. Then for all $x \in (a, b)$, $F'(x) = f(x)$.*

Theorem A.5 (L'Hopital's Rule). *Let f and g be differentiable on (a, b) except possibly at $c \in (a, b)$. Suppose that $\lim_{x \to c} f(x) = 0$, $\lim_{x \to c} g(x) = 0$, $g'(x) \neq 0$ for $x \neq c$, and $\lim_{x \to c} \dfrac{f'(x)}{g'(x)}$ exists. Then*

$$\lim_{x \to c} \frac{f(x)}{g(x)} = \lim_{x \to c} \frac{f'(x)}{g'(x)}.$$

Theorem A.6. *Let x and b be real numbers. Then*

$$\lim_{a \to \infty} \left(1 + \frac{x}{a}\right)^{a+b} = e^x.$$

Theorem A.7 (Some Functions and Their Derivatives). *The following table gives some commonly used functions and their derivatives:*

Function	*Derivative*
x^a $(a \neq 0)$	ax^{a-1}
$\log(x)$ *(or* $\ln x$*)*	$\dfrac{1}{x}$
a^x $(a > 0)$	$a^x \ln a$
e^x *(or* $\exp(x)$*)*	e^x

Theorem A.8 (Binomial Theorem). *Let* $n \geq 1$ *and* $a, b \in \mathbb{R}$. *Then*

$$(a+b)^n = \sum_{i=0}^{n} \binom{n}{i} a^i b^{n-i},$$

where

$$\binom{n}{i} = \frac{n!}{(n-i)!i!} = \frac{n(n-1)\cdots(n-i+1)}{i(i-1)\cdots 1}.$$

Theorem A.9 (Integration by Parts). *Let* f *and* g *be differentiable on* (a, b) *with* $f'g$ *and* fg' *continuous. If* fg' *and* $f'g$ *are integrable over* (a, b), *i.e.,*

$$\int_a^b |f(x)g'(x)| \, \mathrm{d}\, x < \infty, \quad \int_a^b |f'(x)g(x)| \, \mathrm{d}\, x < \infty,$$

then

$$\int_a^b f(x)g'(x) \, \mathrm{d}\, x = f(b)g(b) - f(a)g(a) - \int_a^b f'(x)g(x) \, \mathrm{d}\, x.$$

Theorem A.10 (Integration by Substitution). *Assume that*

- $\phi : [a, b] \to [\phi(a), \phi(b)]$ *is differentiable, and*
- $f : [\phi(a), \phi(b)] \to \mathbb{R}$ *is continuous.*

Then

$$\int_a^b f(\phi(x))\phi'(x) \, \mathrm{d}\, x = \int_{\phi(a)}^{\phi(b)} f(x) \, \mathrm{d}\, x.$$

B

Special Functions

The gamma function and the beta function are commonly encountered in probability and statistics due to the popularity of the beta distribution and the beta distribution. However, the gamma function and the beta function are special functions that are usually not covered in calculus. In this append, we give the definitions and some facts of the two special functions. For further details, readers are referred to [8], [1], and [7].

Definition B.1 (Gamma Function). The gamma function is defined as

$$\Gamma(x) = \int_0^\infty e^{-t} t^{x-1} \, \mathrm{d}\, t, \quad x > 0.$$

Definition B.2 (Beta Function). The beta function is defined as

$$B(x, y) = \int_0^1 t^{x-1}(1-t)^{y-1} \, \mathrm{d}\, t, \quad x > 0, \, y > 0.$$

Theorem B.1 (Gamma Function). *The gamma function* $\Gamma(x)$ *is positive, finite, and*

$$\Gamma(x+1) = x\Gamma(x), \quad x > 0.$$

In addition, $\Gamma(1) = 1$ *and* $\Gamma(\frac{1}{2}) = \sqrt{\pi}$.

Theorem B.2 (Beta Function). *For all* $a > 0$ *and* $b > 0$,

$$B(a, b) = \frac{\Gamma(a)\Gamma(b)}{\Gamma(a+b)}.$$

DOI: 10.1201/9781003484899-B

Theorem B.3 (Stirling's Approximation). *For any real number x,*

$$\lim_{n\to\infty} \Gamma(x+n) = \sqrt{2\pi}e^{-n}n^{x+n-\frac{1}{2}}.$$

In particular, letting $x = 1$ gives the following approximation of $n!$:

$$n! \approx \sqrt{2\pi}e^{-n}n^{n+\frac{1}{2}}.$$

Letting $x = 0$ gives

$$\lim_{\alpha\to\infty} \Gamma(\alpha) = \sqrt{2\pi}e^{-\alpha}\alpha^{\alpha-\frac{1}{2}}.$$

C

Normal Distribution Table

The standard normal random variable Z has the following probability density function:

$$f(x) = \frac{1}{\sqrt{2\pi}} \exp\left(-\frac{x^2}{2}\right), \quad x \in (-\infty, \infty).$$

The cumulative distribution function of Z does not have an explicit formula. It is given by the following integral:

$$\Phi(z) = P(Z \leq z) = \int_{-\infty}^{z} \frac{1}{\sqrt{2\pi}} \exp\left(-\frac{x^2}{2}\right) \mathrm{d}\, x.$$

The standard normal distribution is symmetric about the y-axis. We have the following identity:

$$\Phi(z) + \Phi(-z) = 1, \quad z \in (-\infty, \infty).$$

Table C.2 gives the normal distribution table. It shows the probabilities $P(Z \leq z)$ for $z \in [0, 3.99]$. The value of z to the first decimal is shown in the left column. The second decimal of z is shown in the top row. Values of z for some commonly used values of $\Phi(z)$ are given in Table C.1.

TABLE C.1
Values of z for selected values of $\Phi(z)$.

$\Phi(z)$	0.8	0.85	0.9	0.95	0.975	0.99	0.995
z	0.842	1.036	1.282	1.645	1.960	2.326	2.576

DOI: 10.1201/9781003484899-C

TABLE C.2
The standard normal table.

z	0	0.01	0.02	0.03	0.04	0.05	0.06	0.07	0.08	0.09
0	0.5000	0.5040	0.5080	0.5120	0.5160	0.5199	0.5239	0.5279	0.5319	0.5359
0.1	0.5398	0.5438	0.5478	0.5517	0.5557	0.5596	0.5636	0.5675	0.5714	0.5753
0.2	0.5793	0.5832	0.5871	0.5910	0.5948	0.5987	0.6026	0.6064	0.6103	0.6141
0.3	0.6179	0.6217	0.6255	0.6293	0.6331	0.6368	0.6406	0.6443	0.6480	0.6517
0.4	0.6554	0.6591	0.6628	0.6664	0.6700	0.6736	0.6772	0.6808	0.6844	0.6879
0.5	0.6915	0.6950	0.6985	0.7019	0.7054	0.7088	0.7123	0.7157	0.7190	0.7224
0.6	0.7257	0.7291	0.7324	0.7357	0.7389	0.7422	0.7454	0.7486	0.7517	0.7549
0.7	0.7580	0.7611	0.7642	0.7673	0.7704	0.7734	0.7764	0.7794	0.7823	0.7852
0.8	0.7881	0.7910	0.7939	0.7967	0.7995	0.8023	0.8051	0.8078	0.8106	0.8133
0.9	0.8159	0.8186	0.8212	0.8238	0.8264	0.8289	0.8315	0.8340	0.8365	0.8389
1	0.8413	0.8438	0.8461	0.8485	0.8508	0.8531	0.8554	0.8577	0.8599	0.8621
1.1	0.8643	0.8665	0.8686	0.8708	0.8729	0.8749	0.8770	0.8790	0.8810	0.8830
1.2	0.8849	0.8869	0.8888	0.8907	0.8925	0.8944	0.8962	0.8980	0.8997	0.9015
1.3	0.9032	0.9049	0.9066	0.9082	0.9099	0.9115	0.9131	0.9147	0.9162	0.9177
1.4	0.9192	0.9207	0.9222	0.9236	0.9251	0.9265	0.9279	0.9292	0.9306	0.9319
1.5	0.9332	0.9345	0.9357	0.9370	0.9382	0.9394	0.9406	0.9418	0.9429	0.9441
1.6	0.9452	0.9463	0.9474	0.9484	0.9495	0.9505	0.9515	0.9525	0.9535	0.9545
1.7	0.9554	0.9564	0.9573	0.9582	0.9591	0.9599	0.9608	0.9616	0.9625	0.9633
1.8	0.9641	0.9649	0.9656	0.9664	0.9671	0.9678	0.9686	0.9693	0.9699	0.9706
1.9	0.9713	0.9719	0.9726	0.9732	0.9738	0.9744	0.9750	0.9756	0.9761	0.9767
2	0.9772	0.9778	0.9783	0.9788	0.9793	0.9798	0.9803	0.9808	0.9812	0.9817
2.1	0.9821	0.9826	0.9830	0.9834	0.9838	0.9842	0.9846	0.9850	0.9854	0.9857
2.2	0.9861	0.9864	0.9868	0.9871	0.9875	0.9878	0.9881	0.9884	0.9887	0.9890
2.3	0.9893	0.9896	0.9898	0.9901	0.9904	0.9906	0.9909	0.9911	0.9913	0.9916
2.4	0.9918	0.9920	0.9922	0.9925	0.9927	0.9929	0.9931	0.9932	0.9934	0.9936
2.5	0.9938	0.9940	0.9941	0.9943	0.9945	0.9946	0.9948	0.9949	0.9951	0.9952
2.6	0.9953	0.9955	0.9956	0.9957	0.9959	0.9960	0.9961	0.9962	0.9963	0.9964
2.7	0.9965	0.9966	0.9967	0.9968	0.9969	0.9970	0.9971	0.9972	0.9973	0.9974
2.8	0.9974	0.9975	0.9976	0.9977	0.9977	0.9978	0.9979	0.9979	0.9980	0.9981
2.9	0.9981	0.9982	0.9982	0.9983	0.9984	0.9984	0.9985	0.9985	0.9986	0.9986
3	0.9987	0.9987	0.9987	0.9988	0.9988	0.9989	0.9989	0.9989	0.9990	0.9990
3.1	0.9990	0.9991	0.9991	0.9991	0.9992	0.9992	0.9992	0.9992	0.9993	0.9993
3.2	0.9993	0.9993	0.9994	0.9994	0.9994	0.9994	0.9994	0.9995	0.9995	0.9995
3.3	0.9995	0.9995	0.9995	0.9996	0.9996	0.9996	0.9996	0.9996	0.9996	0.9997
3.4	0.9997	0.9997	0.9997	0.9997	0.9997	0.9997	0.9997	0.9997	0.9997	0.9998
3.5	0.9998	0.9998	0.9998	0.9998	0.9998	0.9998	0.9998	0.9998	0.9998	0.9998
3.6	0.9998	0.9998	0.9999	0.9999	0.9999	0.9999	0.9999	0.9999	0.9999	0.9999
3.7	0.9999	0.9999	0.9999	0.9999	0.9999	0.9999	0.9999	0.9999	0.9999	0.9999
3.8	0.9999	0.9999	0.9999	0.9999	0.9999	0.9999	0.9999	0.9999	0.9999	0.9999
3.9	1.0000	1.0000	1.0000	1.0000	1.0000	1.0000	1.0000	1.0000	1.0000	1.0000

D

R Code

Table D.1 shows a piece of R code used to fit a gamma distribution to the synthetic dataset given in Table 7.1. The `optim` function is used to maximize the log-likelihood function.

TABLE D.1
R code used to fit a gamma distribution to the data.

```
llgamma <- function(param, x) {
  alpha <- param[1]
  theta <- param[2]
  if(alpha <= 0.01 || theta <= 0.01) {
    return(-10^30)
  }
  logdensity <- (alpha-1)*log(x) - x/theta - alpha*log(theta)-
    lgamma(alpha)
  return(sum(logdensity))
}

dat <- c(106,140,146,147,294,337,436,540,566,588,642,
  655,755,762,957,1035,1055,1182,1230,1238,
  1391,1876,2365,2895,4424)

optim(c(2,1500), llgamma, x=dat, control=list(fnscale=-1))
```

DOI: 10.1201/9781003484899-D

E

Solutions to Selected Exercises

E.1 Probability Theory

1.2 By Exercise 1.1, we have

$$P(A \cup B \cup C) = P(A) + P(B \cup C) - P(A \cap (B \cup C))$$

and

$$P(B \cup C) = P(B) + P(C) - P(B \cap C).$$

Note that $A \cap (B \cup C) = (A \cap B) \cup (A \cap C)$. By Exercise 1.1 again, we have

$$\begin{aligned} P(A \cap (B \cup C)) &= P((A \cap B) \cup (A \cap C)) \\ &= P(A \cap B) + P(A \cap C) - P((A \cap B) \cap (A \cap C)) \\ &= P(A \cap B) + P(A \cap C) - P(A \cap B \cap C). \end{aligned}$$

The result follows by combining the above equations.

1.4 By Exercise 1.1, we have

$$P(E \cup F) = P(E) + P(F) - P(E \cap F).$$

Since $P(E \cup F) \leq 1$, we have

$$P(E) + P(F) - P(E \cap F) \leq 1.$$

The result follows by rearranging the equation.

1.6 By Exercise 1.1, we have

$$P(A \cup B) = P(A) + P(B) - P(A \cap B) = 0.7$$

and

$$P(A \cap B^c) = P(A) + P(B^c) - P(A \cap B^c) = 0.9.$$

Adding the above two equations gives

$$1.6 = 2P(A) + P(B) + P(B^c) - P(A \cap B) - P(A \cap B^c).$$

Note that $P(B) + P(B^c) = 1$ and $P(A \cap B) + P(A \cap B^c) = P(A)$. We get

$$1.6 = 2P(A) + 1 - P(A) = P(A) + 1,$$

which gives $P(A) = 0.6$.

DOI: 10.1201/9781003484899-E

1.8 Let A, B, and C denote the events of choosing supplementary coverages A, B, and C, respectively. Then the information gives

$$P(A) = \frac{1}{4}, \quad P(B) = \frac{1}{3}, \quad P(C) = \frac{5}{12}, \quad P(A \cap B \cap C) = 0,$$

$$P(A \cap (B \cup C)^c) = 0, \quad P(B \cap (A \cup C)^c) = 0, \quad P(C \cap (A \cup B)^c) = 0.$$

From the fact that $P(A \cap (B \cup C)^c) = 0$, we get

$$P(A \cap (B \cup C)) = P(A) - P(A \cap (B \cup C)^c) = P(A).$$

Note that

$$P(A \cap (B \cup C)) = P((A \cap B) \cup (A \cap C)) = P(A \cap B) + P(A \cap C) - P(A \cap B \cap C).$$

Combining the above equations leads to

$$P(A \cap B) + P(A \cap C) = P(A) = \frac{1}{4}.$$

Similarly, we have

$$P(B \cap A) + P(B \cap C) = P(B) = \frac{1}{3}, \quad P(C \cap A) + P(C \cap B) = P(C) = \frac{5}{12}.$$

Hence

$$P(A \cap B) + P(B \cap C) + P(C \cap A) = \frac{1}{2}\left(\frac{1}{4} + \frac{1}{3} + \frac{5}{12}\right) = \frac{1}{2}.$$

Then by Exercise 1.2, we get

$$P(A \cup B \cup C) = \frac{1}{4} + \frac{1}{3} + \frac{5}{12} - \frac{1}{2} = \frac{1}{2}.$$

The probability that a randomly selected employee will choose no supplementary coverage is

$$P((A \cup B \cup C)^c) = 1 - P(A \cup B \cup C) = \frac{1}{2}.$$

1.10 Since E and F are independent, we have

$$P(E \cap F) = P(E)P(F).$$

Note that $E \cap F$ and $E \cap F^c$ are disjoint and $(E \cap F) \cup (E \cap F^c) = E$. Then by the third probability axiom, we have

$$P(E) = P(E \cap F) + P(E \cap F^c).$$

Hence

$$\begin{aligned} P(E \cap F^c) &= P(E) - P(E \cap F) = P(E) - P(E)P(F) \\ &= P(E)(1 - P(F)) = P(E)P(F^c), \end{aligned}$$

which shows that E and F^c are independent.

1.12 By the definition of conditional probability, we have

$$P(E_n|E_1 \cap E_2 \cap \cdots \cap E_{n-1}) = \frac{P(E_1 \cap E_2 \cap \cdots \cap E_n)}{P(E_1 \cap E_2 \cap \cdots \cap E_{n-1})},$$

which gives

$$P(E_1 \cap E_2 \cap \cdots \cap E_n) = P(E_n|E_1 \cap E_2 \cap \cdots \cap E_{n-1})P(E_1 \cap E_2 \cap \cdots \cap E_{n-1}).$$

Similarly, we have

$$P(E_1 \cap E_2 \cap \cdots \cap E_{n-1}) = P(E_{n-1}|E_1 \cap E_2 \cap \cdots \cap E_{n-2})P(E_1 \cap E_2 \cap \cdots \cap E_{n-2}).$$

By keeping this process, we get the result.

1.14 Let

$$\begin{aligned} A &= \{\text{A woman has high blood pressure}\}, \\ B &= \{\text{A woman smokes}\}, \\ C &= \{\text{A woman has high blood cholesterol}\}. \end{aligned}$$

Then from the given information, we have

$$P(A \cap (B \cup C)^c) = P(B \cap (A \cup C)^c) = P(C \cap (A \cup B)^c) = 0.1,$$

$$P(A \cap B \cap C^c) = P(B \cap C \cap A^c) = P(A \cap C \cap B^c) = 0.12,$$

and

$$P(A \cap B \cap C|A \cap B) = \frac{1}{3}.$$

Let $x = P(A \cap B \cap C)$. The above information can be visualized in a Venn diagram shown below.

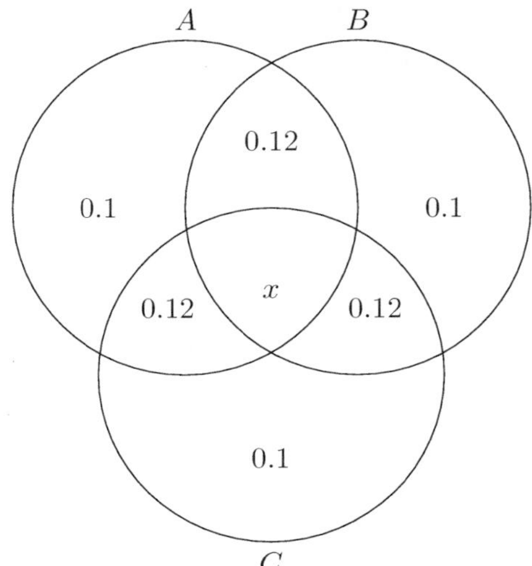

From the last equation, we get

$$\frac{1}{3} = \frac{P(A \cap B \cap C \cap A \cap B)}{P(A \cap B)} = \frac{x}{0.12 + x},$$

which gives $x = 0.06$. We need to calculate $P(A^c \cap B^c \cap C^c | A^c)$. This can be done as follows:

$$\begin{aligned} P(A^c \cap B^c \cap C^c | A^c) = P((A \cup B \cup C)^c | A^c) &= \frac{P((A \cup B \cup C)^c \cap A^c)}{P(A^c)} \\ &= \frac{P((A \cup B \cup C)^c)}{P(A^c)} = \frac{1 - P(A \cup B \cup C)}{1 - P(A)} \\ &= \frac{1 - 0.1 \times 3 - 0.12 \times 3 - 0.06}{1 - 0.1 - 0.12 - 0.12 - 0.06} = \frac{0.28}{0.4} = 0.7. \end{aligned}$$

1.16 The sum of the probabilities is equal to one. That is,

$$a^2 + 0.5a + a + 0.25a + 0.5 = 1,$$

which gives

$$a^2 + 1.75a - 0.5 = 0.$$

Factorizing the left side gives

$$(a + 2)(a - 0.25) = 0.$$

Since $a \geq 0$, we get $a = 0.25$.

1.18 The probability that the loss on a randomly selected claim is greater than the value of the automobile is:

$$P(x > V) = 1 - P(X \leq V) = 1 - F(V) = \frac{1}{10} \exp(0) = 0.1.$$

1.20 The probability that X is odd is calculated as

$$\sum_{k=1}^{\infty} p(2k - 1) = \sum_{k=1}^{\infty} \frac{2}{3^{2k-1}} = 6 \sum_{k=1}^{\infty} \left(\frac{1}{9}\right)^k = 6 \frac{1/9}{1 - 1/9} = \frac{3}{4}.$$

1.22 Let $Y = X^2$. Since X is a random variable with a positive support, we can use Theorem 1.2 to find the pdf of Y. By Theorem 1.2, we have

$$f_Y(y) = \frac{1}{2} y^{-\frac{1}{2}} \frac{1}{\theta} \exp\left(-\frac{\sqrt{y}}{\theta}\right) = \frac{1}{2\theta\sqrt{y}} \exp\left(-\frac{\sqrt{y}}{\theta}\right).$$

1.24 The cdf of Y is

$$\begin{aligned} F_Y(y) =& P(Y \leq y) = P(|X| \leq y) = P(X \leq y) - P(X < -y) \\ =& F_X(y) - F_X(-y). \end{aligned}$$

Hence the pdf of Y is

$$f_Y(y) = f_X(y) + f_X(-y) = 2 f_X(y) = \frac{2}{\sqrt{2\pi}} e^{-\frac{y^2}{2}}, \quad , y \geq 0.$$

1.26 The expected value of X can be calculated as follows:

$$\begin{aligned}E[X] &= \int_{-2}^{4} xf(x)\,\mathrm{d}\,x = \int_{-2}^{4} \frac{x|x|}{10}\,\mathrm{d}\,x \\ &= \int_{-2}^{0} \frac{-x^2}{10}\,\mathrm{d}\,x + \int_{0}^{4} \frac{x^2}{10}\,\mathrm{d}\,x \\ &= \frac{2^3}{30} + \frac{4^3}{30} = 2.4.\end{aligned}$$

1.28 The expected value of X is calculated as

$$\begin{aligned}E[X] &= \int_{0}^{\infty} \max(t,2)f(t)\,\mathrm{d}\,t = \int_{0}^{2} 2f(t)\,\mathrm{d}\,t + \int_{2}^{\infty} tf(t)\,\mathrm{d}\,t \\ &= \int_{0}^{2} \frac{2}{3}e^{-t/3}\,\mathrm{d}\,t + \int_{2}^{\infty} \frac{t}{3}e^{-t/3}\,\mathrm{d}\,t = -2e^{-t/3}\Big|_0^2 - \int_{2}^{\infty} t\,\mathrm{d}\,e^{-t/3}\,\mathrm{d}\,t \\ &= 2 - 2e^{-2/3} - te^{-t/3}\Big|_2^{\infty} + \int_{2}^{\infty} e^{-t/3}\,\mathrm{d}\,t \\ &= 2 + 3e^{-2/3}.\end{aligned}$$

1.30 The distribution function is

$$F(x) = \int_{0}^{x} f(t)\,\mathrm{d}\,t = \int_{0}^{x} \frac{\alpha\theta^{\alpha}}{(t+\theta)^{(\alpha+1)}}\,\mathrm{d}\,t = 1 - \left(\frac{\theta}{x+\theta}\right)^{\alpha}.$$

By the assumptions, we have

$$1 - \left(\frac{\theta}{2\theta - k}\right)^{\alpha} = 0.2, \quad 1 - \left(\frac{\theta}{4\theta - 2k}\right)^{\alpha} = 0.9,$$

which gives $2^{\alpha} = 8$. Hence $\alpha = 3$.

1.32 Since $f(t)$ is continuous, symmetric about $t = 6.5$, and proportional to $1/(t+1)$ between days 0 and 6.5, we have

$$f(t) = \begin{cases} \dfrac{a}{t+1}, & \text{if } 0 \le t < 6.5, \\ \dfrac{a}{14-t}, & \text{if } 6.5 \le t \le 13, \end{cases}$$

where a is a constant to be determined. To determine a, we can use the property of pdfs:

$$\int_{0}^{13} f(t)\,\mathrm{d}\,t = 1,$$

which is

$$\int_{0}^{6.5} \frac{a}{t+1}\,\mathrm{d}\,t + \int_{6.5}^{13} \frac{a}{14-t}\,\mathrm{d}\,t = a\ln 7.5 + a\ln 7.5 = 1.$$

From the above equation, we get $a = \dfrac{1}{2\ln 7.5}$.

Let $\pi_{0.6}$ be the 60th percentile. Then $\pi_{0.6} > 6.5$ since the pdf is symmetric about $t = 6.5$. Hence

$$\begin{aligned} 0.6 =& P(X \le \pi_{0.6}) = 0.5 + \frac{1}{2\ln 7.5} \int_{6.5}^{\pi_{0.6}} \frac{1}{14 - t} \mathrm{d}t \\ =& 0.5 + \frac{1}{2\ln 7.5} \ln \frac{7.5}{14 - \pi_{0.6}}, \end{aligned}$$

which gives $\pi_{0.6} = 8.99$.

1.34 To prove this identity, we can use the properties of the probability measure. Since

$$\begin{aligned} \{X \le a_2, Y \le b_2\} =& \{a_1 < X \le a_2, b_1 < Y \le b_2\} \\ & \cup (\{X \le a_2, Y \le b_1\} \cup \{X \le a_1, Y \le b_2\}) \end{aligned}$$

and

$$\{a_1 < X \le a_2, b_1 < Y \le b_2\} \cap (\{X \le a_2, Y \le b_1\} \cup \{X \le a_1, Y \le b_2\}) = \emptyset,$$

we have

$$\begin{aligned} P(\{X \le a_2, Y \le b_2\}) =& P(\{a_1 < X \le a_2, b_1 < Y \le b_2\}) \\ & + P(\{X \le a_2, Y \le b_1\} \cup \{X \le a_1, Y \le b_2\}). \end{aligned}$$

By Exercise 1.1, we have

$$\begin{aligned} & P(\{X \le a_2, Y \le b_1\} \cup \{X \le a_1, Y \le b_2\}) \\ =& P(\{X \le a_2, Y \le b_1\}) + P(\{X \le a_1, Y \le b_2\}) \\ & - P(\{X \le a_2, Y \le b_1\} \cap \{X \le a_1, Y \le b_2\}) \\ =& P(\{X \le a_2, Y \le b_1\}) + P(\{X \le a_1, Y \le b_2\}) - P(\{X \le a_1, Y \le b_1\}) \\ =& F(a_2, b_1) + F(a_1, b_2) - F(a_1, b_1). \end{aligned}$$

Combining the above equations, we get

$$F(a_2, b_2) = P(\{a_1 < X \le a_2, b_1 < Y \le b_2\}) + F(a_2, b_1) + F(a_1, b_2) - F(a_1, b_1).$$

The result follows by rearranging the above equation.

1.36 To calculate the conditional variance, let us first determine the conditional pf. The conditional pf of Y given $X = 0$ is:

$$p_{Y|X}(0|0) = \frac{p(0,0)}{p_X(0)} = \frac{0.12}{0.12 + 0.06 + 0.05 + 0.02} = \frac{12}{25},$$

$$p_{Y|X}(1|0) = \frac{p(0,1)}{p_X(0)} = \frac{0.06}{0.12 + 0.06 + 0.05 + 0.02} = \frac{6}{25},$$

$$p_{Y|X}(2|0) = \frac{p(0,2)}{p_X(0)} = \frac{0.05}{0.12 + 0.06 + 0.05 + 0.02} = \frac{5}{25},$$

$$p_{Y|X}(3|0) = \frac{p(0,3)}{p_X(0)} = \frac{0.02}{0.12 + 0.06 + 0.05 + 0.02} = \frac{2}{25}.$$

Then we calculate the first two raw moments as follows:

$$\begin{aligned} &E[Y|X=0] \\ =&0 \times p_{Y|X}(0|0) + 1 \times p_{Y|X}(1|0) + 2 \times p_{Y|X}(2|0) + 3 \times p_{Y|X}(3|0) \\ =&\frac{6+10+6}{25} = \frac{22}{25} \end{aligned}$$

$$\begin{aligned} &E[Y^2|X=0] \\ =&0^2 \times p_{Y|X}(0|0) + 1^2 \times p_{Y|X}(1|0) + 2^2 \times p_{Y|X}(2|0) + 3^2 \times p_{Y|X}(3|0) \\ =&\frac{6+20+18}{25} = \frac{44}{25}. \end{aligned}$$

Hence the conditional variance is

$$\text{Var}(Y|X=0) = E[Y^2|X=0] - E[Y|X=0]^2 = \frac{44}{25} - \left(\frac{22}{25}\right)^2 = 0.9856.$$

.38 We first need to figure out $p(1,1)$ from the given information. To do that, we need to calculate the variance of XY. This can done as follows:

$$E[XY] = \sum_{x=0}^{1}\sum_{y=1}^{2} xyp(x,y) = p(1,1) + 2p(1,2) = 7p(1,1),$$

$$E[(XY)^2] = \sum_{x=0}^{1}\sum_{y=1}^{2} (xy)^2 p(x,y) = p(1,1) + 4p(1,2) = 13p(1,1).$$

The variance is

$$\text{Var}(XY) = E[(XY)^2] - E[XY]^2 = 13p(1,1) - 49p(1,1)^2.$$

The variance is maximized when $p(1,1) = \frac{13}{98}$. From this, we get $p(1,2) = p(1,1) = \frac{39}{98}$.

The probability that $X = 0$ or $Y = 0$ can be calculated as

$$\begin{aligned} P(X=0 \text{ or } Y=0) =&1 - P(X \neq 0 \text{ and } Y \neq 0) \\ =&1 - P(X=1, Y=1) - P(X=1, Y=2) \\ =&1 - p(1,1) - p(1,2) = 1 - \frac{13}{98} - \frac{39}{98} = \frac{23}{49}. \end{aligned}$$

1.40 The unconditional probability can be calculated as follows:

$$\begin{aligned} P(N > 2) &= \int_0^5 P(N > 2|\lambda)\frac{1}{5}\,\mathrm{d}\,\lambda \\ &= \frac{1}{5}\int_0^5 \left(1 - e^{-\lambda} - \lambda e^{-\lambda} - \frac{1}{2}\lambda^2 e^{-\lambda}\right)\mathrm{d}\,\lambda \\ &= 0.4344. \end{aligned}$$

1.42 Let X_i be the death benefit paid to the ith employee for $i = 1, 2, \ldots, 1000$. Then

$$P(X_i = 0) = 0.986, \quad P(X_i = 50000) = 0.014.$$

Hence

$$\mu = E[X_i] = 50000 \times 0.014 = 700, \quad E[X_i^2] = 50000^2 \times 0.014 = 3.5 \times 10^7.$$

The standard deviation of X_i is

$$\sigma = \sqrt{E[X_i^2] - E[X_i]^2} = 5874.52.$$

By the central limit theorem, we have

$$P\left(\frac{X_1 + X_2 + \cdots + X_{1000} - 1000\mu}{\sigma\sqrt{1000}} \leq x\right) \approx \Phi(x)$$

Hence the smallest amount of money for the fund is

$$\sigma\sqrt{1000}\Phi^{-1}(0.99) + 1000\mu = 2.326 \times 5874.52 \times \sqrt{1000} + 700000 = 1132098.$$

Rounding the number to the nearest thousand, we get 1,132,000.

1.44 Let N_i be the number of claims filed by the ith policyholder for $i = 1, 2, \ldots, 1250$. By the central limit theorem, we have

$$\frac{N_1 + \cdots + N_{1250} - 1250 \times 2}{\sqrt{2}\sqrt{1250}} \sim N(0, 1).$$

Hence the probability can be approximated by

$$\begin{aligned} &\Phi\left(\frac{2600 - 1250 \times 2}{\sqrt{2}\sqrt{1250}}\right) - \Phi\left(\frac{2450 - 1250 \times 2}{\sqrt{2}\sqrt{1250}}\right) \\ =&\Phi(2) - \Phi(-1) \\ =&\Phi(2) + \Phi(1) - 1 = 0.9772 + 0.8413 - 1 = 0.8185. \end{aligned}$$

E.2 Frequency Models

2.2 By definition, the moment-generating function of N is given by:

$$\begin{aligned} M_N(t) &= E\left[e^{tz}\right] = \sum_{k=0}^{m} e^{kt} p_k \\ &= \sum_{k=0}^{m} e^{kt} \binom{m}{k} q^k (1-q)^{m-k} = \sum_{k=0}^{m} \binom{m}{k} (qe^t)^k (1-q)^{m-k} \\ &= \left[qe^t + 1 - q\right]^m = \left[1 + q(e^t - 1)\right]^m . \end{aligned}$$

2.4 Let J and K denote the number of severe storms in cities J and K, respectively. Then we have

$$p_J(j) = P(J = j) = \binom{5}{j} 0.6^j 0.4^{5-j}, \quad j = 0, 1, \ldots, 5$$

and

$$p_{K|J}(k|j) = P(K = k|J = j) = \begin{cases} \dfrac{1}{2}, & \text{if } k = j, \\ \dfrac{1}{3}, & \text{if } k = j + 1, \\ \dfrac{1}{6}, & \text{if } k = j + 2. \end{cases}$$

We need to determine the conditional distribution $p_{J|K}(j|5)$. By Bayes' theorem (see Exercise 1.9), we have

$$p_{J|K}(j|5) = \frac{p_{K|J}(5|j) p_J(j)}{p_K(5)}, \quad j = 0, 1, \ldots, 5.$$

The above conditional probabilities are non-zero only at $j = 3, 4, 5$:

$$p_{K|J}(5|3) p_J(3) = \frac{1}{6} \binom{5}{3} 0.6^3 0.4^2 = 0.0575,$$

$$p_{K|J}(5|4) p_J(4) = \frac{1}{3} \binom{5}{4} 0.6^4 0.4^1 = 0.0864,$$

$$p_{K|J}(5|5) p_J(5) = \frac{1}{2} \binom{5}{5} 0.6^5 0.4^0 = 0.03888.$$

In addition,

$$p_K(5) = \sum_{j=0}^{5} p_{K|J}(5|j) p_J(j) = 0.18278.$$

Hence

$$p_{J|K}(j|5) = \begin{cases} 0.3151, & \text{if } j = 3, \\ 0.4727, & \text{if } j = 4, \\ 0.2127, & \text{if } j = 5. \end{cases}$$

The conditional expectation is calculated as

$$E[J|K = 5] = \sum_{j=3}^{4} p_{J|K}(j|5) = 3.8996.$$

2.6 Let C be the random variable that indicates the types of the distribution of N. Then we have

$$\begin{aligned} P(N = 2) =& P(N = 2|C = 1)P(C = 1) + P(N = 2|C = 2)P(C = 2) \\ =& p\binom{2}{2}0.5^2 + (1 - p)\binom{4}{2}0.5^4 = 0.375 - 0.125p. \end{aligned}$$

2.8 By the definition of probability generating functions, we have

$$\begin{aligned} M(t) = E\left[e^{tN}\right] &= \sum_{k=0}^{\infty} e^{tk} \frac{\lambda^k e^{-\lambda}}{k!} \\ &= e^{-\lambda} \sum_{k=0}^{\infty} \frac{(e^t\lambda)^k}{k!} \\ &= e^{-\lambda} e^{e^t\lambda} = e^{\lambda(e^t - 1)}. \end{aligned}$$

2.10 We rewrite the function in the left side as follows:

$$\begin{aligned} \binom{m}{k} q^k (1-q)^{m-k} &= \frac{m!}{k!(m-k)!} q^k (1-q)^{m-k} \\ &= \frac{1}{k!} \cdot \frac{m!}{e^{-m} m^{m+\frac{1}{2}} \sqrt{2\pi}} \cdot \frac{e^{-m+k}(m-k)^{m-k+\frac{1}{2}}\sqrt{2\pi}}{(m-k)!} \cdot \\ &\quad \frac{e^{-m} m^{m+\frac{1}{2}} \sqrt{2\pi}}{e^{-m+k}(m-k)^{m-k+\frac{1}{2}}\sqrt{2\pi}} q^k (1-q)^{m-k}. \end{aligned}$$

By Theorem B.3, we have

$$\lim_{m\to\infty} \frac{m!}{e^{-m} m^{m+\frac{1}{2}} \sqrt{2\pi}} = 1$$

and

$$\lim_{m\to\infty} \frac{e^{-m+k}(m-k)^{m-k+\frac{1}{2}}\sqrt{2\pi}}{(m-k)!} = 1.$$

In addition, by Theorems A.6 and B.3, we have

$$\begin{aligned}
&\frac{e^{-m}m^{m+\frac{1}{2}}\sqrt{2\pi}}{e^{-m+k}(m-k)^{m-k+\frac{1}{2}}\sqrt{2\pi}}q^k(1-q)^{m-k}\\
&=\frac{(m-k)^k q^k(1-q)^{m-k}}{e^k\left(1-\frac{k}{m}\right)^{m+\frac{1}{2}}}\\
&=\frac{(mq-kq)^k\left(1-\frac{mq}{m}\right)^{m-k}}{e^k\left(1-\frac{k}{m}\right)^{m+\frac{1}{2}}}\\
&\to\frac{\lambda^k e^{-\lambda}}{e^k e^{-k}} \quad \text{as } m\to\infty, q\to 0, mq\to\lambda\\
&=\lambda^k e^{-\lambda}.
\end{aligned}$$

The result follows by combining the above equations.

2.12 Since N follows a compound distribution, we can use the tower property (see Theorem 1.10) to calculate the mean and the variance. The mean is:

$$E[N]=E[E[N|\Lambda]]=E[\Lambda]=\int_0^2\lambda\frac{1}{2}\,\mathrm{d}\,\lambda=1.$$

To calculate the variance, we first calculate the second raw moment:

$$E[N^2]=E[E[N^2|\Lambda]]=E[\Lambda^2+\Lambda]=\int_0^2(\lambda^2+\lambda)\frac{1}{2}\,\mathrm{d}\,\lambda=\frac{7}{3}.$$

Hence the variance is

$$\mathrm{Var}(N)=E[N^2]-E[N]^2=\frac{4}{3}.$$

2.14 By Definition 1.21, the moment generating function is

$$\begin{aligned}
M(t)=&E\left[e^{tN}\right]=\sum_{k=0}^{\infty}e^{tk}\frac{\Gamma(k+r)}{\Gamma(k+1)\Gamma(r)}p^r(1-p)^k\\
=&p^r\sum_{k=0}^{\infty}\frac{\Gamma(k+r)}{\Gamma(k+1)\Gamma(r)}((1-p)e^t)^k\\
=&p^r\left[1-(1-p)e^t\right]^{-r}.
\end{aligned}$$

2.16 Let E be the event that the PhD program has to make more than 5 offers. Then E^c is the event that the PhD program only needs to make at most 5 offers. We have

$$P(E^c)=\binom{4}{2}0.6^3 0.4^2+\binom{3}{1}0.6^3 0.4+\binom{2}{0}0.6^3 0.4^0=0.6826,$$

which gives

$$P(E)=1-P(E^c)=0.3174.$$

2.18 Let $\{p_k^M\}_{k\geq 0}$ and $P^M(z)$ be the probability function and the probability generating function of the zero-modified distribution, respectively. Then by Theorem 2.5, we have

$$\begin{aligned} P^M(z) =& \sum_{k=0}^{\infty} p_k^M z^k = p_0^M + \sum_{k=1}^{\infty} \frac{1-p_0^M}{1-p_0} p_k z^k \\ =& p_0^M + \frac{1-p_0^M}{1-p_0}(P(z) - p_0) = \frac{p_0^M - p_0}{1-p_0} + \frac{1-p_0^M}{1-p_0} P(z). \end{aligned}$$

2.20 By Theorem 2.5, we have

$$p_k^M = \frac{1-p_0^M}{1-p_0} p_k, \quad k = 1, 2, \ldots.$$

Hence

$$\begin{aligned} p_1^M &= \frac{1-0.5}{1-e^{-3}} \frac{3^1}{1!} e^{-3} = 0.0786. \\ p_2^M &= \frac{1-0.5}{1-e^{-3}} \frac{3^2}{2!} e^{-3} = 0.1179. \\ p_3^M &= \frac{1-0.5}{1-e^{-3}} \frac{3^3}{3!} e^{-3} = 0.1179. \end{aligned}$$

The pgf is

$$P^M(t) = \frac{p_0^M - p_0}{1-p_0} + \frac{1-p_0^M}{1-p_0} e^{\lambda(t-1)}.$$

Then

$$\begin{aligned} E[N] = P^{M'}(1) = \frac{1-p_0^M}{1-p_0}\lambda = 1.5786. \\ E[N(N-1)] = P^{M''}(1) = \frac{1-p_0^M}{1-p_0}\lambda^2 = 4.7358. \end{aligned}$$

The variance is $\mathrm{Var}(N) = E[N^2] - E[N]^2 = 3.8224$.

2.22 From the recursive formula, we have

$$p_k = \frac{2^k}{k!} p_0, \quad k = 1, 2, \ldots.$$

Since the sum of all probabilities is equal to 1, we have

$$1 = \sum_{k=0}^{\infty} p_k = p_0 \sum_{k=0}^{\infty} \frac{2^k}{k!} = p_0 e^2,$$

which gives $p_0 = e^{-2}$. Hence

$$p_4 = \frac{2^4}{4!} e^{-2} = 0.09.$$

E.3 Severity Models

3.2 From Example 3.1, we have $q(\mu) = \sqrt{2\pi}\sigma \exp\left(\frac{\mu^2}{2\sigma^2}\right)$ and $r(\mu) = \frac{\mu}{\sigma^2}$. Taking derivatives of $r(\mu)$ and $q(\mu)$ with respect to μ, we get

$$r'(\mu) = \frac{1}{\sigma^2},$$

$$q'(\mu) = \sqrt{2\pi}\sigma \exp\left(\frac{\mu^2}{2\sigma^2}\right)\frac{\mu}{\sigma^2} = q(\theta)\frac{\mu}{\sigma^2}.$$

By Theorem 3.1, we get

$$E[X] = \frac{q'(\mu)}{r'(\mu)q(\mu)} = \mu.$$

In addition, we have

$$\mathrm{Var}(X) = \frac{\frac{\partial \mu}{\partial \mu}}{r'(\mu)} = \frac{1}{1/\sigma^2} = \sigma^2.$$

3.4 Since $q(\theta)$ is a normalizing factor, we have

$$q(\theta) = \int_0^\infty x e^{-\theta x} \,\mathrm{d}\, x.$$

By integration by parts, we get

$$q(\theta) = \int_0^\infty -\frac{x}{\theta} \,\mathrm{d}\, e^{-\theta x} = -\frac{x}{\theta} e^{-\theta x}\Big|_0^\infty + \frac{1}{\theta}\int_0^\infty e^{-\theta x}\,\mathrm{d}\, x = \frac{1}{\theta^2}.$$

3.6 The mean of X is calculated as

$$E[X] = \int_0^\infty x f(x)\,\mathrm{d}\, x = \int_0^\infty \frac{\tau (x/\theta)^{\alpha\tau} e^{-(x/\theta)^\tau}}{\Gamma(\alpha)}\,\mathrm{d}\, x.$$

To calculate the above integral, we let $y = (x/\theta)^\tau$. Then

$$x = \theta y^{1/\tau}$$

and

$$\mathrm{d}\, x = \frac{\theta}{\tau} y^{1/\tau - 1}\,\mathrm{d}\, y.$$

By the above change of variables, we get

$$\begin{aligned} E[X] &= \frac{1}{\Gamma(\alpha)}\int_0^\infty \tau y^\alpha e^{-y} \frac{\theta}{\tau} y^{1/\tau-1}\,\mathrm{d}\, y = \frac{\theta}{\Gamma(\alpha)}\int_0^\infty y^{\alpha+1/\tau-1} e^{-y}\,\mathrm{d}\, y \\ &= \frac{\Gamma(\alpha + 1/\tau)\theta}{\Gamma(\alpha)}. \end{aligned}$$

Similarly, we calculate the second raw moment of X as follows:

$$E[X^2] = = \frac{\theta^2}{\Gamma(\alpha)} \int_0^\infty y^{\alpha+2/\tau-1} e^{-y} \, \mathrm{d}\, y = \frac{\Gamma(\alpha+2/\tau)\theta^2}{\Gamma(\alpha)}.$$

Hence the variance of X is

$$\mathrm{Var}(X) = E[X^2] - E[X]^2 = \frac{\Gamma(\alpha+2/\tau)\theta^2}{\Gamma(\alpha)} - \left(\frac{\Gamma(\alpha+1/\tau)\theta}{\Gamma(\alpha)}\right)^2$$

3.8 Let $t = x - 1$. By this change of variable, we have

$$\begin{aligned}
\int_1^\infty x^n e^{-x} \, \mathrm{d}\, x = \int_0^\infty (t+1)^n e^{-t-1} \, \mathrm{d}\, t = e^{-1} \int_0^\infty \sum_{j=0}^n \binom{n}{j} t^j e^{-t} \, \mathrm{d}\, t \\
= e^{-1} \sum_{j=0}^n \binom{n}{j} \int_0^\infty t^j e^{-t} \, \mathrm{d}\, t = e^{-1} \sum_{j=0}^n \binom{n}{j} \Gamma(j+1) \\
= e^{-1} \sum_{j=0}^n \frac{n!}{j!(n-j)!} j! = e^{-1} \sum_{j=0}^n \frac{n!}{(n-j)!} \\
= e^{-1} \sum_{j=0}^n \frac{n!}{j!}.
\end{aligned}$$

3.10 This is the gamma function. By the properties of the gamma function, we have

$$\int_0^\infty e^{-t} t^{5/2} \, \mathrm{d}\, t = \Gamma\left(\frac{7}{2}\right) = \frac{5}{2}\Gamma\left(\frac{5}{2}\right) = \frac{5}{2} \cdot \frac{3}{2} \cdot \frac{1}{2}\Gamma\left(\frac{1}{2}\right) = \frac{15}{8}\sqrt{\pi}.$$

3.12 Since GB2 random variables are continuous, the mode of X can be obtained by equating the derivative of the density function to zero and solving the equation. To do that, we rewrite the density as follows:

$$f(x) = \frac{\gamma}{B(\alpha,\tau)\theta^{\gamma\tau}} x^{\gamma\tau-1} \left(1 + x^\gamma \theta^{-\gamma}\right)^{-\alpha-\tau}.$$

By using the chain rule, we get the following derivative of $f(x)$:

$$\begin{aligned}
f'(x) = & \frac{\gamma}{B(\alpha,\tau)\theta^{\gamma\tau}} (\gamma\tau - 1) x^{\gamma\tau-2} \left(1 + x^\gamma \theta^{-\gamma}\right)^{-\alpha-\tau} + \\
& \frac{\gamma}{B(\alpha,\tau)\theta^{\gamma\tau}} x^{\gamma\tau-1} (-\alpha-\tau) \left(1 + x^\gamma \theta^{-\gamma}\right)^{-\alpha-\tau-1} \theta^{-\gamma} \gamma x^{\gamma-1} \\
= & \frac{\gamma x^{\gamma\tau-2} \left(1 + x^\gamma \theta^{-\gamma}\right)^{-\alpha-\tau-1}}{B(\alpha,\tau)\theta^{\gamma\tau}} \left((\gamma\tau-1)\left(1 + x^\gamma\theta^{-\gamma}\right) - \gamma(\alpha+\tau) x^\gamma \theta^{-\gamma}\right)
\end{aligned}$$

Letting $f'(x) = 0$ and ignoring the common factor, we get the following equation:

$$\gamma\tau - 1 - (\gamma\alpha + 1) x^\gamma \theta^{-\gamma} = 0.$$

Solving the above equation, we get

$$x_{mode} = \theta \left(\frac{\gamma\tau - 1}{\gamma\alpha + 1} \right)^{\frac{1}{\gamma}}.$$

3.14 Let $Y = X^{-1}$. By Theorem 1.2, we have

$$\begin{aligned} f_Y(y) = y^{-2} f_X\left(y^{-1}\right) &= y^{-2} \frac{\Gamma(\alpha+\tau)}{\Gamma(\alpha)\Gamma(\tau)} \frac{\gamma (y^{-1}/\theta)^{\gamma\tau}}{y^{-1}[1 + (y^{-1}/\theta)^{\gamma}]^{\alpha+\tau}} \\ &= \frac{\Gamma(\alpha+\tau)}{\Gamma(\alpha)\Gamma(\tau)} \frac{\gamma\left(y/\theta^{-1}\right)^{-\gamma\tau}}{y[1 + (y/\theta^{-1})^{-\gamma}]^{\alpha+\tau}}, \end{aligned}$$

from which we see that $Y \sim GB2\left(\alpha, \frac{1}{\theta}, -\gamma, \tau\right)$.

3.16 By the properties of the beta function and the gamma function, we can calculate the first raw moment as

$$\begin{aligned} E[X] &= \int_0^1 x f(x) \, \mathrm{d}\, x = \int_0^1 \frac{x^{\alpha}(1-x)^{\beta-1}}{B(\alpha,\beta)} = \frac{B(\alpha+1,\beta)}{B(\alpha,\beta)} \\ &= \frac{\Gamma(\alpha+1)\Gamma(\beta)}{\Gamma(\alpha+1+\beta)} \cdot \frac{\Gamma(\alpha+\beta)}{\Gamma(\alpha)\Gamma(\beta)} = \frac{\alpha}{\alpha+\beta}. \end{aligned}$$

Similarly, the second raw moment of X is

$$\begin{aligned} E[X^2] &= \int_0^1 x^2 f(x) \, \mathrm{d}\, x = \int_0^1 \frac{x^{\alpha+1}(1-x)^{\beta-1}}{B(\alpha,\beta)} = \frac{B(\alpha+2,\beta)}{B(\alpha,\beta)} \\ &= \frac{\Gamma(\alpha+2)\Gamma(\beta)}{\Gamma(\alpha+2+\beta)} \cdot \frac{\Gamma(\alpha+\beta)}{\Gamma(\alpha)\Gamma(\beta)} = \frac{(\alpha+1)\alpha}{(\alpha+\beta+1)(\alpha+\beta)}. \end{aligned}$$

The variance of X is

$$\begin{aligned} \mathrm{Var}(X) &= E[X^2] - E[X]^2 = \frac{(\alpha+1)\alpha}{(\alpha+\beta+1)(\alpha+\beta)} - \frac{\alpha^2}{(\alpha+\beta)^2} \\ &= \frac{(\alpha+1)\alpha(\alpha+\beta) - \alpha^2(\alpha+\beta+1)}{(\alpha+\beta+1)(\alpha+\beta)^2} \\ &= \frac{\alpha\beta}{(\alpha+\beta)^2(\alpha+\beta+1)}. \end{aligned}$$

3.18 Let $t = \dfrac{x}{1+x}$. By this change of variable, we get

$$\int_0^{\infty} x^6 (1+x)^{-14} \, \mathrm{d}\, x = \int_0^1 t^6 (1-t)^6 \, \mathrm{d}\, t.$$

By the properties of the beta function and the gamma function, we have

$$\int_0^1 t^6(1-t)^6 \, \mathrm{d}\, t = B(7,7) = \frac{\Gamma(7)\Gamma(7)}{\Gamma(14)} = \frac{6!6!}{13!} = \frac{1}{12012}.$$

3.20 The pdf of the Pareto distribution is

$$f(x) = \frac{2 \times 500000^2}{(x+500000)^3}, \quad x > 0.$$

The expected value of the bonus is

$$\begin{aligned}
&E[B]\\
=&0.15 \int_0^{480000} (480000 - x)\frac{2 \times 500000^2}{(x+500000)^3}\,\mathrm{d}\,x\\
=&0.15 \int_0^{480000} 980000\frac{2 \times 500000^2}{(x+500000)^3}\,\mathrm{d}\,x - 0.15\int_0^{480000} \frac{2 \times 500000^2}{(x+500000)^2}\,\mathrm{d}\,x\\
=&0.15 \times 980000 \left(\frac{-500000^2}{(x+500000)^2}\Bigg|_0^{480000} \right) - 0.15\left(\frac{-2 \times 500000^2}{x+500000}\Bigg|_0^{480000} \right)\\
=&0.15 \times 980000 \left(1 - \frac{500000^2}{980000^2}\right) - 0.15\left(2 \times 500000 - \frac{2 \times 500000^2}{980000}\right)\\
=&108734.69 - 73469.39 = 35265.3.
\end{aligned}$$

E.4 Aggregate Loss Models

4.2 By the property of the moment generating function (see Theorem 1.5), we have

$$\begin{aligned}
E[S] =&M_S'(0) = P_N'(M_X(0))M_X'(0) = P_N'(1)M_X'(0)\\
=&E[N]E[X]
\end{aligned}$$

and

$$\begin{aligned}
E[S^2] =&M_S''(0) = P_N''(M_X(0))M_X'(0)^2 + P_N'(M_X(0))M_X''(0)\\
=&E[N(N-1)]E[X]^2 + E[N]E[X^2]\\
=&E[N]\operatorname{Var}(X) + E[N^2]E[X]^2.
\end{aligned}$$

The variance can be obtained by

$$\operatorname{Var}(S) = E[S^2] - E[S]^2 = E[N]\operatorname{Var}(X) + \operatorname{Var}(N)E[X]^2.$$

4.4 Let first find the 2-fold convolution:

$$\begin{aligned}
f_S^{*2}(x) =&\sum_{y=0}^{x} f_X(x-y)f_X(y) = \sum_{y=0}^{x} p(1-p)^{x-y}\cdot p(1-p)^y\\
=&\binom{x+1}{1}(x+1)p^2(1-p)^x, \quad x = 0,1,2,\ldots.
\end{aligned}$$

The 3-fold convolution is:

$$
\begin{aligned}
f_S^{*3}(x) =& \sum_{y=0}^{x} f_X^{*2}(x-y) f_X(y) = \sum_{y=0}^{x} (x-y+1)p^2(1-p)^{x-y} \cdot p(1-p)^y \\
=& p^3(1-p)^x \sum_{y=0}^{x} (x-y+1) \\
=& \binom{x+2}{2} p^3 (1-p)^x, \quad x = 0, 1, 2, \ldots.
\end{aligned}
$$

Continuing the process, we get

$$
f_S^{*n} = \binom{x+n-1}{n-1} p^n (1-p)^x, \quad x = 0, 1, 2, \ldots.
$$

4.6 Since N follows the Poisson distribution with parameter λ, we have

$$
\operatorname{Var}(N) = E[N] = \lambda.
$$

By Theorem 4.2, the variance of S is

$$
\begin{aligned}
\operatorname{Var}(S) =& E[N] \operatorname{Var}(X) + \operatorname{Var}(N) E[X]^2 = E[N]\left(\operatorname{Var}(X) + E[X]^2\right) \\
=& E[N] E\left[X^2\right].
\end{aligned}
$$

4.8 The pdf of X is

$$
f_X(x) = \frac{1}{\theta} e^{-\frac{x}{\theta}}, \quad x > 0.
$$

The pdf of Y is $f_Y(x) = 1$ for $x \in [0, 1]$. The pdf of S is the convolution of $f_X(x)$ and $f_Y(x)$, i.e.,

$$
\begin{aligned}
f_S(x) =& \int_0^x f_X(x-y) f_Y(y) \, \mathrm{d}y = \begin{cases} \int_0^x \frac{1}{\theta} e^{-\frac{x-y}{\theta}} \, \mathrm{d}y, & \text{if } 0 < x \le 1, \\ \int_0^1 \frac{1}{\theta} e^{-\frac{x-y}{\theta}} \, \mathrm{d}y, & \text{if } 1 < x, \end{cases} \\
=& \begin{cases} 1 - e^{-\frac{x}{\theta}}, & \text{if } 0 < x \le 1, \\ e^{-\frac{x}{\theta}} \left(e^{\frac{1}{\theta}} - 1\right), & \text{if } 1 < x. \end{cases}
\end{aligned}
$$

4.10 Suppose that S is approximately normally distributed with parameters μ and σ^2. Then

$$
\hat{\mu} = 2000, \quad \hat{\sigma} = \sqrt{\operatorname{Var}(S)} = 10000.
$$

Hence

$$
\begin{aligned}
P(S > 3000) =& P\left(\frac{S - 2000}{10000} > \frac{3000 - 2000}{10000}\right) = 1 - \Phi(0.1) \\
=& 1 - 0.5398 = 0.4602.
\end{aligned}
$$

E.5 Coverage Modifications

5.2 By the definition of $e_X(d)$ and the fact that $f_X(x) = -S'(x)$, we have

$$\begin{aligned} e_X(d) =& \frac{-\int_d^\infty (x-d)\,\mathrm{d}\,S(x)}{1-F(d)} = \frac{-(x-d)S(x)|_d^\infty + \int_d^\infty S(x)\,\mathrm{d}\,x}{S(d)} \\ =& \frac{\int_d^\infty S(x)\,\mathrm{d}\,x}{S(d)}. \end{aligned}$$

5.4 Since $F(d) < 1$, we know that $e_X^k(d)$ is defined. By the definition, we have

$$E\left[(X-d)_+^k\right] = \int_d^\infty (x-d)^k f_X(x)\,\mathrm{d}\,x = e_X^k(d)(1-F(d)).$$

5.6 First let us calculate the mean:

$$\begin{aligned} E[X] =& \int_0^{120} x f_X(x)\,\mathrm{d}\,x = \int_0^{80} 0.01x\,\mathrm{d}\,x + \int_{80}^{120} \frac{120x - x^2}{4000}\,\mathrm{d}\,x \\ =& \left.\frac{0.01x^2}{2}\right|_0^{80} + \left.\frac{60x^2 - x^3/3}{4000}\right|_{80}^{120} \\ =& 32 + 18.67 = 50.67. \end{aligned}$$

Then we calculate $E[X \wedge 20]$:

$$\begin{aligned} E[X \wedge 20] =& \int_0^{120} \min(x, 20) f_X(x)\,\mathrm{d}\,x \\ =& \int_0^{20} 0.01x\,\mathrm{d}\,x + \int_{20}^{80} 0.01 \times 20\,\mathrm{d}\,x + \int_{80}^{120} \frac{120-x}{4000} \times 20\,\mathrm{d}\,x \\ =& 2 + 12 + 4 = 18. \end{aligned}$$

The loss elimination ratio is

$$\frac{E[X \wedge d]}{E[X]} = \frac{18}{50.67} = 0.36.$$

5.8 Let d be the ordinary deductible. By Exercise 5.3, we know that the loss elimination ratio is

$$\frac{E[X \wedge d]}{E[X]} = 1 - \left(\frac{\theta}{d+\theta}\right)^{\alpha-1} = 1 - \frac{1000}{d+1000} = 0.7.$$

Solving the above equation gives

$$d = 2333.33.$$

5.10 By definition, we have

$$\begin{aligned}
E[X \wedge u] &= \int_0^\infty \min(x, u) f_X(x)\, \mathrm{d}\, x = \int_0^\infty \min(x, u) \frac{1}{\theta} \exp\left(-\frac{x}{\theta}\right) \mathrm{d}\, x \\
&= \int_0^u \frac{x}{\theta} \exp\left(-\frac{x}{\theta}\right) \mathrm{d}\, x + \int_u^\infty \frac{u}{\theta} \exp\left(-\frac{x}{\theta}\right) \mathrm{d}\, x \\
&= \int_0^u -x\, \mathrm{d} \exp\left(-\frac{x}{\theta}\right) - u \exp\left(-\frac{x}{\theta}\right)\Big|_d^\infty = \int_0^u \exp\left(-\frac{x}{\theta}\right) \mathrm{d}\, x \\
&= \theta\left(1 - \exp\left(-\frac{u}{\theta}\right)\right).
\end{aligned}$$

5.12 Let X denote the claim amount. Then the payment is $Y = \min(X, 1000)$. The expected payment is

$$\begin{aligned}
E[Y] &= E[\min(X, 1000)] = \int_0^{1000} x\, \mathrm{d}\, F(x) + \int_{1000}^\infty 1000\, \mathrm{d}\, F(x) \\
&= xF(x)|_0^{1000} - \int_0^{1000} F(x)\, \mathrm{d}\, x + 1000(1 - F(1000)) \\
&= 1000 - \int_0^{1000} \left(1 - 0.8e^{-0.02x} - 0.2e^{-0.001x}\right) \mathrm{d}\, x \\
&= 1000 - \left(x + 40e^{-0.02x} + 200e^{-0.001x}\right)\Big|_0^{1000} \\
&= 166.42.
\end{aligned}$$

5.14 Let X be the loss random variable. Let Y be the amount paid by the insurance plan. Then

$$Y = \begin{cases} 0, & \text{if } X < 250, \\ 0.75(X - 250), & \text{if } 250 \le X < 2250, \\ 1500, & \text{if } 2250 \le X < 5100, \\ 1500 + 0.95(X - 5100), & \text{if } X \ge 5100, \end{cases}$$

where the bound value 5100 is derived from the given information:

$$250 + 0.25 * (2250 - 250) + X - 2250 = 3600.$$

We can show that

$$Y = 0.75(X \wedge 2250 - X \wedge 250) + 0.95(X - X \wedge 5100).$$

Hence

$$E[Y] = 0.75(E[X \wedge 2250] - E[X \wedge 250]) + 0.95(E[X] - E[X \wedge 5100]).$$

By Exercise 5.3, we get

$$\begin{aligned}
E[Y] &= 0.75\left(\frac{1000 \times 2250}{2250 + 1000} - \frac{1000 \times 250}{250 + 1000}\right) + 0.95\left(1000 - \frac{1000 \times 5100}{5100 + 1000}\right) \\
&= 369.23 + 155.74 = 524.97.
\end{aligned}$$

5.16 By Theorem 5.3, the mean of the per-payment variable is

$$E[Y^P] = \frac{c\,(E[X \wedge u] - E[X \wedge d])}{1 - F(d)}.$$

From the given information, we have $c = 0.8$ and $d = 20$. The policy limit u can be derived as follows:

$$0.8(u - 20) = 60,$$

which gives $u = 95$. Then we can calculate the components as follows:

$$E[X \wedge u] = \int_0^{100} \min(x, u)\frac{x}{5000}\,\mathrm{d}\,x = u - \frac{u^3}{30000},$$
$$F(d) = \int_0^{d} \frac{x}{5000}\,\mathrm{d}\,x = \frac{d^2}{10000},$$

which give

$$E[X \wedge 95] = 66.42, \quad E[X \wedge 20] = 19.73, \quad F(20) = 0.04.$$

Hence

$$E[Y^P] = \frac{0.8\,(66.42 - 19.73)}{1 - 0.04} = 38.91.$$

5.18 By definition, we have

$$\begin{aligned} E[(S - (j+1)h)_+] &= \sum_{k=j+1}^{\infty} (kh - (j+1)h)P(S = kh) \\ &= \sum_{k=j+1}^{\infty} (kh - jh)P(S = kh) - \sum_{k=j+1}^{\infty} hP(S = kh) \\ &= \sum_{k=j}^{\infty}(kh - jh)P(S = kh) - h(1 - P(S \le jh)) \\ &= E[(S - jh)_+] - h(1 - F_S(jh)). \end{aligned}$$

5.20 From the given information, we know that the support of S is $\{0, 1, 2, \ldots\}$. By Exercise 5.18, we have

$$\begin{aligned} E[(S-2)_+] &= E[(S-1)_+] - (1 - F_S(1)) = E[S] - (1 - F_S(0)) - (1 - F_S(1)) \\ &= E[S] - 2 + 2P(S = 0) + P(S = 1). \end{aligned}$$

The mean of S is

$$E[S] = E[N]E[X] = 2 \times 2 = 4.$$

The probabilities of S at 0 and 1 are

$$P(S = 0) = P(N = 0) = e^{-2}, \quad P(S = 1) = P(N = 1)P(X = 1) = \frac{2e^{-2}}{3}.$$

Hence

$$E[(S-2)_+] = 4 - 2 + 2e^{-2} + \frac{2e^{-2}}{3} = 2.36.$$

5.22 By definition, we have

$$E[(S-d)_+] = \int_d^\infty (1 - F(x))\,\mathrm{d}\,x.$$

Hence we have

$$\int_{100}^{120} [1 - F(x)]\,\mathrm{d}\,x = 15 - 10 = 5.$$

Since $P(80 < S \leq 120) = 0$, we have $F(x) = F(80)$ for $x \in (80, 120)$. Hence

$$20[1 - F(80)] = 5,$$

which gives $F(80) = 0.75$.

E.6 Model Estimation

6.2 The mean of the exponential distribution is

$$E[X] = \theta.$$

Hence

$$\hat{\theta} = \frac{98 + 399 + 1067 + 1273 + 1366}{5} = 840.6.$$

6.4 The pdf of the exponential distribution is $f(x) = \frac{1}{\theta}\exp\left(-\frac{x}{\theta}\right)$. Given the n observations, the likelihood function is

$$L(\theta) = \prod_{i=1}^n \frac{1}{\theta}\exp(-\frac{x_i}{\theta}) = \theta^{-n}\exp\left(-\frac{1}{\theta}\sum_{i=1}^n x_i\right) = \theta^{-n}\exp\left(-\frac{n\bar{x}}{\theta}\right).$$

The log-likelihood function is

$$l(\theta) = -n\ln\theta - \frac{n\bar{x}}{\theta}.$$

Solving

$$l'(\theta) = -\frac{n}{\theta} + \frac{n\bar{x}}{\theta^2} = 0,$$

we get $\hat{\theta} = \bar{x}$.

6.6 The likelihood function is

$$\begin{aligned} L(\alpha) =& (\alpha+1)0.74^\alpha \cdot (\alpha+1)0.81^\alpha \cdot (\alpha+1)0.95^\alpha \\ =& (\alpha+1)^3 0.56943^\alpha. \end{aligned}$$

The log-likelihood function is

$$l(\alpha) = 3\ln(\alpha+1) + \alpha \ln 0.56943.$$

Solving

$$l'(\theta) = \frac{3}{\alpha+1} + \ln 0.56943 = 0,$$

we get $\hat{\alpha} = 4.3275$.

6.8 The density function of the Pareto distribution is

$$f(x) = \frac{\alpha 70^\alpha}{(x+70)^{\alpha+1}}.$$

The log-likelihood function is

$$l(\alpha) = 3\ln\alpha + 3\alpha\ln 70 - (\alpha+1)\sum_{i=1}^{3}\ln(x_i+70).$$

Solving $l'(\alpha) = 0$ gives

$$\hat{\alpha} = \frac{3}{\sum_{i=1}^{3}\ln\frac{x_i+70}{70}} = 3.0022.$$

6.10 The likelihood function is the product of the probabilities of the four events:

$$\begin{aligned} L(\theta) =& \frac{1}{2\theta}\exp\left(-\frac{1}{2\theta}\right) \cdot \frac{1}{2\theta}\exp\left(-\frac{2}{2\theta}\right) \cdot \frac{1}{2\theta}\exp\left(-\frac{3}{2\theta}\right) \cdot \frac{1}{3\theta}\exp\left(-\frac{15}{3\theta}\right) \\ =& \frac{1}{24\theta^4}\exp\left(-\frac{8}{\theta}\right). \end{aligned}$$

The log-likelihood function is

$$l(\theta) = -\ln 24 - 4\ln\theta - \frac{8}{\theta}.$$

Solving

$$l'(\theta) = -\frac{4}{\theta} + \frac{8}{\theta^2} = 0,$$

we get $\hat{\theta} = 2$.

6.12 We need to use the conditional probability for losses with deductibles. For the Pareto distribution, we have

$$P(X = x) = \frac{\alpha\theta^\alpha}{(x+\theta)^{\alpha+1}}, \quad P(X > x) = \frac{\theta^\alpha}{(x+\theta)^\alpha}.$$

The likelihood function is given by

$$\begin{aligned}
&L(\alpha)\\
=&P(X = 200|X > 100)^3 P(X = 300)^2 P(X > 2000)^3 P(X = 400|X > 200)^2\\
=&\left(\frac{\alpha(100+\theta)^\alpha}{(200+\theta)^{\alpha+1}}\right)^3 \left(\frac{\alpha\theta^\alpha}{(300+\theta)^{\alpha+1}}\right)^2 \left(\frac{\theta^\alpha}{(2000+\theta)^{\alpha+1}}\right)^3 \left(\frac{\alpha(200+\theta)^\alpha}{(400+\theta)^{\alpha+1}}\right)^2\\
=&\frac{\alpha^7\theta^{5\alpha}(100+\theta)^{3\alpha}}{(200+\theta)^{\alpha+3}(300+\theta)^{2\alpha+2}(400+\theta)^{2\alpha+2}(2000+\theta)^{3\alpha+3}} =
\end{aligned}$$

The log-likelihood function is

$$l(\alpha) = 7\ln\alpha + \alpha\ln\frac{\theta^5(100+\theta)^3}{(200+\theta)(300+\theta)^2(400+\theta)^2(2000+\theta)^3} + c,$$

where c is a constant that is independent of α. Taking the derivative of $l(\alpha)$, we get

$$l'(\alpha) = \frac{7}{\alpha} + \ln\frac{\theta^5(100+\theta)^3}{(200+\theta)(300+\theta)^2(400+\theta)^2(2000+\theta)^3}.$$

Solving $l'(\alpha) = 0$ with $\theta = 1000$ gives

$$\hat{\alpha} = 1.5946.$$

6.14 The log-likelihood function of the sample is

$$\begin{aligned}
l(\lambda) =&\ln\left(\prod_{i=1}^{10}\frac{\lambda^{x_i}e^{-\lambda}}{x_i!}\right) = \sum_{i=1}^{10}(x_i\ln\lambda - \lambda - \ln(x_i!))\\
=&39\ln\lambda - 10\lambda - \sum_{i=1}^{10}\ln(x_i!).
\end{aligned}$$

The first derivative and the second derivative of $l(\lambda)$ are given by

$$\begin{aligned}
l'(\lambda) =&\frac{39}{\lambda} - 10,\\
l''(\lambda) =&-\frac{39}{\lambda^2}.
\end{aligned}$$

The maximum likelihood estimate of λ is $\hat{\lambda} = 3.9$. The Fisher information is

$$I(\lambda) = -E\left[-\frac{39}{\lambda^2}\right] = \frac{39}{\lambda^2}.$$

The asymptotic variance of the estimate is

$$I(\hat{\lambda})^{-1} = \frac{\hat{\lambda}^2}{39} = 0.39.$$

6.16 Let Y be the new losses from the portfolio. Then $E[Y|\Theta] = \dfrac{1}{2}\theta$. By Equation (6.6), we can calculate the Bayes estimate of Y as follows:

$$\begin{aligned} E[Y|\mathbf{x}] &= \int_{600}^{\infty} E[Y|\theta] f_{\Theta|X}(\theta|\mathbf{x})\, \mathrm{d}\,\theta = \int_{600}^{\infty} \frac{1}{2}\theta \frac{3\times 600^3}{\theta^4}\, \mathrm{d}\,\theta \\ &= \frac{3\times 600^3}{2}\int_{600}^{\infty}\theta^{-3}\, \mathrm{d}\,\theta = \frac{3\times 600}{4} = 450. \end{aligned}$$

6.18 Let $\mathbf{x} = \{10\}$ be the observed data. The joint distribution of N and the parameter Θ is

$$f(n,\lambda) = f(\lambda)P(N=n|\Theta=\lambda) = \left(\frac{1}{15}\exp\left(-\frac{\lambda}{6}\right) + \frac{1}{20}\exp\left(-\frac{\lambda}{12}\right)\right)\frac{\lambda^n e^{-\lambda}}{n!}$$

Hence

$$\begin{aligned} f(\mathbf{x},\lambda) &= \left(\frac{1}{15}\exp\left(-\frac{\lambda}{6}\right) + \frac{1}{20}\exp\left(-\frac{\lambda}{12}\right)\right)\frac{\lambda^{10} e^{-\lambda}}{10!} \\ &= \frac{1}{10!}\left(\frac{\lambda^{10}\exp\left(-\frac{7}{6}\lambda\right)}{15} + \frac{\lambda^{10}\exp\left(-\frac{13}{12}\lambda\right)}{20}\right). \end{aligned}$$

The posterior distribution of Θ is

$$f_{\Theta|N}(\lambda|\mathbf{x}) = \frac{f(\mathbf{x},\lambda)}{\int_0^\infty f(\mathbf{x},u)\,\mathrm{d}\,u} = \frac{\dfrac{\lambda^{10}\exp\left(-\frac{7}{6}\lambda\right)}{15} + \dfrac{\lambda^{10}\exp\left(-\frac{13}{12}\lambda\right)}{20}}{\int_0^\infty \dfrac{u^{10}\exp\left(-\frac{7}{6}u\right)}{15} + \dfrac{u^{10}\exp\left(-\frac{13}{12}u\right)}{20}\,\mathrm{d}\,u}.$$

Let Y be the number of claims from this policyholder in Year 2. Given $\Theta = \lambda$ the mean of Y is $E[Y|\Theta=\lambda] = \lambda$. By Equation (6.6), we have

$$\begin{aligned} E[Y|\mathbf{x}] &= \int_0^\infty \frac{\dfrac{\lambda^{11}\exp\left(-\frac{7}{6}\lambda\right)}{15} + \dfrac{\lambda^{11}\exp\left(-\frac{13}{12}\lambda\right)}{20}}{\int_0^\infty \dfrac{u^{10}\exp\left(-\frac{7}{6}u\right)}{15} + \dfrac{u^{10}\exp\left(-\frac{13}{12}u\right)}{20}\,\mathrm{d}\,u}\,\mathrm{d}\,\lambda \\ &= \frac{\int_0^\infty \dfrac{\lambda^{11}\exp\left(-\frac{7}{6}\lambda\right)}{15} + \dfrac{\lambda^{11}\exp\left(-\frac{13}{12}\lambda\right)}{20}\,\mathrm{d}\,\lambda}{\int_0^\infty \dfrac{u^{10}\exp\left(-\frac{7}{6}u\right)}{15} + \dfrac{u^{10}\exp\left(-\frac{13}{12}u\right)}{20}\,\mathrm{d}\,u} \end{aligned}$$

$$= \frac{\frac{1}{15}\left(\frac{6}{7}\right)^{12}\int_0^\infty t^{11}e^{-t}\,\mathrm{d}\,t + \frac{1}{20}\left(\frac{12}{13}\right)^{12}\int_0^\infty t^{11}e^{-t}\,\mathrm{d}\,t}{\frac{1}{15}\left(\frac{6}{7}\right)^{11}\int_0^\infty t^{10}e^{-t}\,\mathrm{d}\,t + \frac{1}{20}\left(\frac{12}{13}\right)^{11}\int_0^\infty t^{10}e^{-t}\,\mathrm{d}\,t}$$

$$= \frac{\frac{1}{15}\left(\frac{6}{7}\right)^{12}\Gamma(12) + \frac{1}{20}\left(\frac{12}{13}\right)^{12}\Gamma(12)}{\frac{1}{15}\left(\frac{6}{7}\right)^{11}\Gamma(11) + \frac{1}{20}\left(\frac{12}{13}\right)^{11}\Gamma(11)} = 9.88.$$

E.7 Model Selection

7.2 The cdf of X is

$$F(x) = 1 - \exp\left(-\frac{x}{1000}\right).$$

The quantile at p is

$$Q(p) = \inf\{x : F(x) \geq p\} = \inf\left\{x : 1 - \exp\left(-\frac{x}{1000}\right) \geq p\right\}$$
$$= \inf\{x : x \geq -1000\ln(1-p)\} = -1000\ln(1-p).$$

When $p = 0.15$, we get $Q(0.15) = 162.5189$.

7.4 For a P-P plot, the coordinates are $(F_n(x_j), F(x_j))$. Hence the coordinates for $x_7 = 120$ are

$$\left(\frac{7}{20+1}, 1 - \exp(-120/427.5)\right) = (0.3333, 0.2447).$$

7.6 Let $x_0 = 0$ and $x_{n+1} = \infty$. Then we have

$$\max_{0\leq x\leq\infty} |F_n(x) - F(x)| = \max_{0\leq i\leq n}\ \max_{x_i\leq x<x_{i+1}} |F_n(x) - F(x)|.$$

The empirical cdf is constant in the intervals $[x_i, x_{i+1})$. The cdf of the distribution is nondecreasing. As a result, there are three cases regarding the interaction of $F_n(x)$ and $F(x)$ in $[x_i, x_{i+1})$: $F_n(x)$ is above $F(x)$, $F_n(x)$ is below $F(x)$, and $F_n(x)$ intersects $F(x)$. In the first case, we have for all $x \in [X_i, x_{i+1})$,

$$|F_n(x) - F(x)| = F_n(x) - F(x) \leq F_n(x_i) - F(x_i).$$

In the second case, we have all $x \in [x_i, x_{i+1})$,

$$|F_n(x) - F(x)| = F(x) - F_n(x) \leq F(x_{i+1}) - F_n(x_{i+1}-).$$

In the third case, let $F_n(x)$ and $F(x)$ intersect at $x_* \in [x_i, x_{i+1})$. Then $F_n(x)$ is above $F(x)$ in $[x_i, x_*]$and $F_n(x)$ is below $F(x)$ in $[x_*, x_{i+1})$. Based on the previous analysis, we have

$$|F_n(x) - F(x)| \leq \max(F_n(x_i) - F(x_i), F(x_{i+1}) - F_n(x_{i+1}-)).$$

Hence all $x \in [x_i, x_{i+1})$,

$$|F_n(x) - F(x)| \leq \max(|F_n(x_i) - F(x_i)|, |F(x_{i+1}) - F_n(x_{i+1}-)|),$$

which gives

$$\max_{x_i \leq x < x_{i+1}} |F_n(x) - F(x)| = \max(|F_n(x_i) - F(x_i)|, |F(x_{i+1}) - F_n(x_{i+1}-)|).$$

Hence

$$\begin{aligned}
&\max_{0\leq i\leq n} \max_{x_i \leq x < x_{i+1}} |F_n(x) - F(x)| \\
&= \max_{0\leq i\leq n} \max(|F_n(x_i) - F(x_i)|, |F(x_{i+1}) - F_n(x_{i+1}-)|) \\
&= \max_{1\leq i\leq n} \max(|F_n(x_i) - F(x_i)|, |F(x_i) - F_n(x_i-)|).
\end{aligned}$$

The last step follows from the fact that $|F_n(x_0) - F(x_0)| = 0$ and $|F_n(x_{n+1}) - F(x_{n+1})| = 0$.

7.8 Since $F(x) = 1 - e^{-x/\theta}$, we have $F(200) = 1 - \exp(-200/427.5) = 0.3736$. Note that $F_n(200) = 10/20 = 0.5$. We have

$$D(200) = |F_n(x) - F(x)| = |0.5 - 0.3736| = 0.1264.$$

7.10 Let E_i and O_i be the expected and observed number of claims in the ith group for $i = 1, 2, 3$. Then we can calculate the chi-square goodness-of-fit test statistic as follows:

	E_i	O_i	$\frac{(E_i - O_i)^2}{E_i}$
A	117.992	112	0.3043
B	151.016	180	5.5628
C	160.992	138	3.2836

Adding the individual terms gives

$$C^2 = 9.1507.$$

7.12 The test statistic is

$$T = 2\ln(L_1/L_0) = 2(\ln L_1 - \ln L_0) = 2\ln 3 = 2.1972.$$

Since the statistic follows a χ^2 distribution with 2 degrees of freedom, the critical value is 4.605. Since $2.1972 < 4.605$, we fail to reject the null hypothesis.

7.14 The BICs for the three models are given by

$$\begin{aligned} AIC_A =&(\ln 200) \times 1 - 2 \times (-205) = 415.2983, \\ AIC_B =&(\ln 200) \times 2 - 2 \times (-203) = 416.5966, \\ AIC_C =&(\ln 200) \times 3 - 2 \times (-200) = 415.8950. \end{aligned}$$

Since the first model has the lowest BIC, the first model is favored by the BIC.

7.16 By the definition of the AIC and the BIC, we have

$$\begin{aligned} 2 \times 3 - 2l(\hat{\theta}) =&606, \\ 2 \ln n - 2l(\hat{\theta}) =&612.4292. \end{aligned}$$

Solving the above equations, we get $n = 500$.

7.18 The ratios for this dataset are given by

k	1	2	3	4	5	6	7	8
$k\frac{n_k}{n_{k-1}}$	1.10	1.73	1.52	3.64	2.50	3.20	3.50	4.00

Plotting the ratios against k, we see that the points form a line that is approximately straight and trends up. This indicates that the negative binomial model is a suitable model for the data.

E.8 Credibility Models

8.2 Let X_1, X_2, …, X_n be the observations of the aggregate losses. The jth aggregate loss is

$$X_j = Y_{j1} + Y_{j2} + \cdots + Y_{jN_j},$$

where N_j is the number of claims. Since the claim severity follows the Pareto distribution (see Exercise 3.15) and the claim frequency follows the negative binomial distribution, we have

$$\begin{aligned} E[N] =&\frac{r(1-p)}{p} = 0.4, \\ \mathrm{Var}(N) =&\frac{r(1-p)}{p^2} = 0.48, \\ E[Y] =&\frac{\theta}{\alpha - 1} = 500, \\ \mathrm{Var}(Y) =&\frac{\alpha\theta^2}{(\alpha-1)^2(\alpha-2)} = 3 \times 500^2. \end{aligned}$$

By Theorem 4.2, we have

$$\begin{aligned}\mu =& E[X_j] = E[Y]E[N] = 0.4 \times 500 = 200,\\ \sigma =& \sqrt{\operatorname{Var}(X_j)} = \sqrt{E[N]\operatorname{Var}(Y) + \operatorname{Var}(N)E[Y]^2}\\ =& \sqrt{0.4 \times 3 \times 500^2 + 0.48 \times 500^2} = 500\sqrt{1.68}.\end{aligned}$$

By Theorem 8.1, the full credibility requirement is

$$n \geq \lambda_0 \left(\frac{\sigma}{\mu}\right)^2 = \left(\frac{\Phi^{-1}((1+0.9)/2)}{0.05}\right)^2 \left(\frac{500\sqrt{1.68}}{200}\right)^2 = 11363.92.$$

The minimum number of observations required for full credibility is 11364.

8.4 Let X denote the claim severity random variable. Note that X follows the Pareto distribution with parameters $\alpha = 6$ and $\theta = 0.5$. By Exercise 3.15, we have

$$\mu_X = E[X] = \frac{\theta}{\alpha - 1} = \frac{0.5}{6-1} = 0.1,$$

$$\sigma_X^2 = \operatorname{Var}(X) = \frac{\alpha\theta^2}{(\alpha-1)^2(\alpha-2)} = 0.015.$$

By Theorem 8.2, we have

$$\lambda \geq \left(\frac{\Phi^{-1}(0.95)}{0.02}\right)\left(1 + \frac{0.015}{0.1^2}\right) = 16913.$$

8.6 Let $S = X_1 + X_2 + \cdots + X_N$ be the aggregate losses. Let Θ be the distribution of the parameter θ. By the tower properties of the conditional expectation, we have

$$\begin{aligned}E[S|\Theta] =& E[E[S|N,\Theta]|\Theta] = E[NE[X|\Theta]|\Theta] = E[N|\Theta] \cdot E[X|\Theta]\\ =& \Theta \cdot 10\Theta = 10\Theta^2.\end{aligned}$$

Similarly, we have

$$\begin{aligned}E[S^2|\Theta] =& E[E[S^2|N,\Theta]|\Theta] = E[NE[X^2|\Theta] + N(N-1)E[X|\Theta]^2|\Theta]\\ =& E[N|\Theta] \cdot E[X^2|\Theta] + E[N(N-1)|\Theta] \cdot E[X|\Theta]^2\\ =& \Theta \cdot 2(10\Theta)^2 + \Theta^2 \cdot (10\Theta)^2 = 100\Theta^4 + 200\Theta^3.\end{aligned}$$

From the above equations, we get

$$\operatorname{Var}(S|\Theta) = E[S^2|\Theta] - E[S|\Theta]^2 = 200\Theta^3.$$

Then we can calculate the expectation of the process variance and the variance of the hypothetical mean as follows:

$$\begin{aligned}\mu_{PV} =& E[\operatorname{Var}(S|\Theta)] = E[200\Theta^3] = \int_1^\infty 200\theta^3 \cdot 5\theta^{-6}\, \mathrm{d}\,\theta\\ =& -500\theta^{-2}\big|_1^\infty = 500,\end{aligned}$$

$$\begin{aligned}\sigma_{HM}^2 &= \text{Var}(E[S|\Theta]) = \text{Var}(10\Theta^2) = 100E[\Theta^4] - 100E[\Theta^2]^2\\ &= 100\int_1^{\infty} \theta^4 \cdot 5\theta^{-6}\,\mathrm{d}\,\theta - 100\left(\int_1^{\infty} \theta^2 \cdot 5\theta^{-6}\,\mathrm{d}\,\theta\right)^2 = \frac{2000}{9}.\end{aligned}$$

Hence

$$k = \frac{\mu_{PV}}{\sigma_{HM}^2} = \frac{500}{2000/9} = \frac{9}{4} = 2.25.$$

8.8 To apply the Bühlmann credibility model, we need to calculate $\mu_{PV} = E[\text{Var}(X|\Lambda)]$ and $\sigma_{HM}^2 = \text{Var}(E[X|\Lambda])$, where X is the number of claims in a year. By the assumption, we can calculate those quantities as follows:

$$\begin{aligned}\mu_{PV} &= E[\text{Var}(X|\Lambda)] = E[\Lambda] = \alpha\theta = 1.2,\\ \sigma_{HM}^2 &= \text{Var}(E[X|\Lambda]) = \text{Var}(\Lambda) = \alpha\theta^2 = 1.44,\end{aligned}$$

$$E[X] = E[E[X|\Lambda]] = E[\Lambda] = 1.2.$$

Hence

$$k = \frac{\mu_{PV}}{\sigma_{HM}^2} = \frac{1.2}{1.44} = 0.8333.$$

The Bühlmann credibility estimate of the number of claims in Year 3 is

$$\frac{k}{2+k}E[X] + \frac{2}{2+k}\bar{X} = \frac{0.8333}{2.8333} \cdot 1.2 + \frac{2}{2.8333} \cdot \frac{3+0}{2} = 1.41.$$

8.10 From the joint distribution, we can get the marginal distribution for Θ as follows:

$$P(\Theta = 0) = \sum_{x=0}^{2} P(X = x, \Theta = 0) = 0.6,$$

$$P(\Theta = 1) = \sum_{x=0}^{2} P(X = x, \Theta = 1) = 0.4.$$

Let $\mu(\Theta) = E[X|\Theta]$ and $\sigma^2(\Theta) = \text{Var}(X|\Theta)$. From the given information, we have

$$\mu(\theta) = \sum_{x=0}^{2} x p_{X|\Theta}(x|\theta) = \sum_{x=0}^{2} x\frac{P(X = x, \Theta = \theta)}{P(\Theta = \theta)} = \begin{cases} 0.5, & \text{if } \theta = 0,\\ 1, & \text{if } \theta = 1,\end{cases}$$

and

$$\begin{aligned}\sigma^2(\theta) &= E[X^2|\Theta = \theta] - \mu(\theta)^2 = \sum_{x=0}^{2} x^2\frac{P(X = x, \Theta = \theta)}{P(\Theta = \theta)} - \mu(\theta)^2\\ &= \begin{cases} 7/12, & \text{if } \theta = 0,\\ 1/2, & \text{if } \theta = 1.\end{cases}\end{aligned}$$

Then we can calculate k as follows:

$$k = \frac{E[\sigma^2(\Theta)]}{\mathrm{Var}(\mu(\Theta))} = \frac{\frac{7}{12}\cdot 0.6 + \frac{1}{2}\cdot 0.4}{0.5^2\cdot 0.6 + 1^2\cdot 0.4 - (0.5\cdot 0.6 + 1\cdot 0.4)^2} = \frac{55}{6}.$$

The Bühlmann credibility premium is

$$\frac{k}{n+k}E[\mu(\Theta)] + \frac{n}{n+k}\bar{X} = \frac{55/6}{10+55/6}\cdot 0.7 + \frac{10}{10+55/6}\cdot 1 = 0.8565.$$

8.12 Let X denote the loss per employee. From the given information, we have

$$E[X] = 20, \quad \mu_{PV} = 8000, \quad \sigma^2_{HM} = 40,$$

$$\bar{X}_{selected} = \frac{800\times 15 + 600\times 10 + 400\times 5}{800+600+400} = 11.1111.$$

By Theorem 8.4, we have

$$k = \frac{\mu_{PV}}{\sigma^2_{HM}} = \frac{8000}{40} = 200.$$

The Bühlmann-Straub credibility premium per employee for the selected policy is

$$\begin{aligned}&\frac{k}{m_1+m_2+m_3+k}E[X] + \frac{m_1+m_2+m_3}{m_1+m_2+m_3+k}\bar{X}_{selected}\\ =&\frac{200\times 20 + 1800\times 11.111}{800+600+400+200} = 12.\end{aligned}$$

8.14 Let us first determine the posterior distribution $p_{\Theta|X}(q|\mathbf{x})$, where $\mathbf{x} = (1,1)$. By the assumption, we have

$$p_{X|\Theta}(\mathbf{x}|q) = 2q(1-q)\cdot 2q(1-q) = 4q^2(1-q)^2.$$

Hence

$$\begin{aligned}p_{\Theta|X}(q|\mathbf{x}) =&\frac{p_{X|\Theta}(\mathbf{x}|q)f_\Theta(q)}{\int_0^1 p_{X|\Theta}(\mathbf{x}|s)f_\Theta(s)\,\mathrm{d}\,s} = \frac{16q^5(1-q)^2}{\int_0^1 16s^5(1-s)^2\,\mathrm{d}\,s}\\ =&\frac{q^5(1-q)^2}{B(6,3)} = \frac{\Gamma(9)}{\Gamma(6)\Gamma(3)}q^5(1-q)^2 = 168q^5(1-q)^2.\end{aligned}$$

Then the Bayes estimate of N is

$$\begin{aligned}E[E[N|\Theta]|\mathbf{x}] =& \int_0^1 E[N|\Theta = q]\cdot p_{\Theta|X}(q|\mathbf{x})\,\mathrm{d}\,q = \int_0^1 2q\cdot 168q^5(1-q)^2\,\mathrm{d}\,q\\ =&2\cdot 168B(7,3) = 2\cdot 168\frac{\Gamma(6)\Gamma(2)}{\Gamma(10)} = \frac{4}{3}.\end{aligned}$$

8.16 We treat each policy as a risk group. Then we have $r = 4$ and $n_i = 7$ for $i = 1, 2, 3, 4$. By Theorem 8.6 and the given information, we have

$$\tilde{\mu}_{PV} = \frac{33.6}{4 \times (7-1)} = 1.4$$

and

$$\tilde{\sigma}^2_{HM} = \frac{7 \times 3.3 - (4-1) \times 1.4}{(4-1)7} = 0.9.$$

Hence we have

$$k = \frac{\tilde{\mu}_{PV}}{\tilde{\sigma}^2_{HM}} = \frac{1.4}{0.9} = \frac{14}{9}.$$

The Bühlmann credibility factor for an individual policy is

$$\frac{n_i}{n_i + k} = \frac{7}{7 + 14/9} = \frac{9}{11}.$$

E.9 Risk Measures

9.2 Let X be a uniform random variable on $[0, 2]$ and le Y be a uniform random variable on $[2, 3]$. Then $Y \geq X$. The risk measures of X and Y are

$$\rho(X) = \sqrt{E[X^2] - E[X]^2} = \sqrt{\frac{1}{3}} = \frac{\sqrt{3}}{3}$$

and

$$\rho(Y) = \sqrt{E[Y^2] - E[Y]^2} = \frac{\sqrt{3}}{6}.$$

Since $\rho(Y) < \rho(X)$, this risk measure does not have the monotonicity property.

9.4 Let $Y = cX$. Since $c > 0$,we have

$$F_Y(x) = P(Y \leq x) = P(cX \leq x) = P(X \leq c^{-1}x) = F_X(c^{-1}x).$$

Hence

$$\begin{aligned}
\mathrm{VaR}_p(cX) &= \inf\{x : F_Y(x) \geq p\} = \inf\{x : F_X(c^{-1}x) \geq p\} \\
&= \inf\{cx : F_X(x) \geq p\} = c \inf\{x : F_X(x) \geq p\} \\
&= c\,\mathrm{VaR}_p(X).
\end{aligned}$$

9.6 Let us first find the cdf of X:

$$\begin{aligned}
F_X(x) =& P(X \leq x) = P(U + T \leq x) \\
=& P(U \leq x)P(T = 0) + P(U + 1 \leq x)P(T = 1) \\
=& 0.99P(U \leq x) + 0.01P(U \leq x - 1).
\end{aligned}$$

Since $\mathrm{VaR}_{0.95}(X) = \inf\{x : F_X(x) \geq 0.95\}$, we need to find x such that

$$0.99P(U \leq x) + 0.01P(U \leq x - 1) = 0.95.$$

Since $0 \leq U \leq 1$, the value of x must be less than 1. Otherwise, $P(U \leq x) = 1$, which makes the above equation impossible. When $x < 1$, the above equation becomes

$$0.99P(X \leq x) = 0.95,$$

which gives $x = 95/99 = 0.9596$.

9.8 The cdf of X is:

$$\begin{aligned}
F(x) = \int_0^x f(u)\,\mathrm{d}\,u &= \int_0^x \frac{\beta}{1000} \exp\left(-\frac{u}{1000}\right) \mathrm{d}\,u + \int_0^x \frac{1-\beta}{500} \exp\left(-\frac{u}{500}\right) \mathrm{d}\,u \\
&= \beta\left(1 - \exp\left(-\frac{x}{1000}\right)\right) + (1-\beta)\left(1 - \exp\left(-\frac{x}{500}\right)\right) \\
&= 1 - \beta \exp\left(-\frac{x}{1000}\right) - (1-\beta)\exp\left(-\frac{x}{500}\right).
\end{aligned}$$

To calculate the VaR at 0.9, we need to solve $F(x) = 0.9$, which is

$$1 - \beta \exp\left(-\frac{x}{1000}\right) - (1-\beta)\exp\left(-\frac{x}{500}\right) = 0.9.$$

Let $y = \exp\left(-\frac{x}{1000}\right)$. Then the above equation becomes

$$(1-\beta)y^2 + \beta y - 0.1 = 0,$$

which gives

$$y = \frac{-\beta + \sqrt{\beta^2 - 0.4\beta + 0.4}}{2(1-\beta)}.$$

Hence

$$\mathrm{VaR}_{0.9}(X) = -1000 \ln \frac{-\beta + \sqrt{\beta^2 - 0.4\beta + 0.4}}{2(1-\beta)}.$$

9.10

(a) By definition, we have

$$E[X^k] = \int_a^b x^k \frac{1}{b-a}\,\mathrm{d}\,x = \frac{b^{k+1} - a^{k+1}}{(k+1)(b-a)}.$$

(b) $\mathrm{Var}(X) = E[X^2] - E[X]^2 = \dfrac{(b-a)^2}{12}$

(c) By definition, we have

$$e(d) = \frac{\int_d^b (x-d)\frac{1}{b-a}\,\mathrm{d}\,x}{1 - \frac{d-a}{b-a}} = \frac{b-d}{2}.$$

(d) The cdf is

$$F(x) = \int_a^x f(t)\,\mathrm{d}\,t = \frac{x-a}{b-a}.$$

Hence

$$\mathrm{VaR}_p = F^{-1}(p) = p(b-a) + a.$$

(e) Since $\mathrm{TVaR}_p = \mathrm{VaR}_p + e(\pi_p)$, we have

$$\mathrm{TVaR}_p = p(b-a) + a + \frac{b - p(b-a) - a}{2} = \frac{(b-a)(1_+p)}{2} + a.$$

9.12 We first need to determine the cdf of S. The cdf of S can be calculated as follows:

$$\begin{aligned} F_S(x) =& P(S \le x) = P(X+Y \le x) = \int_{0<s<1,0<t<1,s+t<x} 2\,\mathrm{d}\,s\,\mathrm{d}\,t \\ =& \int_0^x \int_0^{x-s} 2\,\mathrm{d}\,t\,\mathrm{d}\,s = \int_0^x 2(x-s)\,\mathrm{d}\,s \\ =& x^2, \quad 0 < x < 1. \end{aligned}$$

Hence

$$\mathrm{VaR}_{0.75}(S) = F_S^{-1}(0.75) = \frac{\sqrt{3}}{2}.$$

The TVaR of S at level 0.75 is

$$\begin{aligned} \mathrm{TVaR}_{0.75}(S) =& E[S|S > \mathrm{VaR}_{0.75}(S)] \\ =& \frac{1}{1-0.75} \int_{\sqrt{3}/2}^1 x \cdot 2x\,\mathrm{d}\,x = \frac{8 - 3\sqrt{3}}{3}. \end{aligned}$$

9.14 The cdf of X is

$$F(x) = \int_0^x f(x)\,\mathrm{d}\,x = \frac{1000x - 0.5x^2}{500000}, \quad x \in (0, 1000).$$

The VaR at 0.75 is

$$\mathrm{VaR}_{0.75}(X) = F^{-1}(0.75) = 500.$$

Hence

$$\mathrm{TVaR}_{0.75}(X) = \frac{1}{1-0.75} \int_{500}^{1000} \frac{x(1000-x)}{500000} = 666.67.$$

Bibliography

[1] George E. Andrews, Richard Askey, and Ranjan Roy. *Special Functions.* Cambridge University Press, Cambridge, UK, 1999.

[2] Philippe Artzner, Freddy Delbaen, Jean-Marc Eber, and David Heath. Coherent measures of risk. *Mathematical Finance*, 9(3):203–228, July 1999.

[3] David B. Atkinson. Credibility methods applied to life, health, and pensions, 2019. `https://www.soa.org/globalassets/assets/files/resources/tables-calcs-tools/credibility-methods-life-health-pensions.pdf`. Accessed on February 12, 2024.

[4] Hans Bühlmann. Experience rating and credibility. *ASTIN Bulletin*, 5(2):157–165, May 1969.

[5] J.David Cummins, Georges Dionne, James B. McDonald, and B. Michael Pritchett. Applications of the GB2 family of distributions in modeling insurance loss processes. *Insurance: Mathematics and Economics*, 9(4):257–272, 1990.

[6] Guojun Gan and Emiliano A Valdez. Regression modeling for the valuation of large variable annuity portfolios. *North American Actuarial Journal*, 22(1):40–54, 2018.

[7] Sudhir R. Ghorpade and Balmohan V. Limaye. *A Course in Calculus and Real Analysis.* Springer, New York, NY, 2018.

[8] Omar Hijab. *Introduction to Calculus and Classical Analysis.* Undergraduate Texts in Mathematics. Springer, New York, NY, 4th edition, 2016.

[9] Wu-Yi Hsiang. *A Concise Introduction to Calculus.* World Scientific Publishing Company, Singapore, 1995.

[10] Rob J. Hyndman and Yanan Fan. Sample quantiles in statistical packages. *The American Statistician*, 50(4):361, November 1996.

[11] Norman L. Johnson, Adrienne W. Kemp, and Samuel Kotz. *Univariate discrete distributions.* Wiley, Hoboken, NJ, 3rd edition, 2005.

[12] Christian Kleiber and Samuel Kotz. *Statistical size distributions in economics and actuarial sciences.* Wiley, Hoboken, NJ, 2003.

[13] Stuart A. Klugman, Harry H. Panjer, and Gordon E. Willmot. *Loss models: From Data to Decisions.* Wiley series in probability and statistics. Wiley, Hoboken, NJ, 5th edition, 2019.

[14] J. F. Lawless. Inference in the generalized gamma and log gamma distributions. *Technometrics*, 22(3):409–419, aug 1980.

[15] Amita Majumder and Satya Ranjan Chakravarty. Distribution of personal income: Development of a new model and its application to U.S. income data. *Journal of Applied Econometrics*, 5(2):189–196, 1990.

[16] James B. McDonald. Some generalized functions for the size distribution of income. *Econometrica*, 52(3):647, may 1984.

[17] Sheldon M. Ross. *A First Course in Probability.* Pearson, Harlow, UK, 10th edition, 2020.

[18] Jiafeng Sun, Edward W. Frees, and Marjorie A. Rosenberg. Heavy-tailed longitudinal data modeling using copulas. *Insurance: Mathematics and Economics*, 42(2):817–830, 2008.

[19] Gary G Venter. Transformed beta and gamma distributions and aggregate losses. *Proceedings of the Casualty Actuarial Society*, 70:156–193, 1983.

[20] David Williams. *Probability with martingales.* Cambridge mathematical textbooks. Cambridge University Press, Cambridge, UK, 1991.

List of Symbols

Index

Printed in the United States